Exploring Textual Data

Text, Speech and Language Technology

VOLUME 4

The titles published in this series are listed at the end of this volume.

Exploring Textual Data

by

Ludovic Lebart
Centre National de la Recherche Scientifique, Paris

André Salem
Université de la Sorbonne Nouvelle, Paris

and

Lisette Berry
L. Berry Associates, Inc., New York

KLUWER ACADEMIC PUBLISHERS
DORDRECHT / BOSTON / LONDON

A C.I.P. Catalogue record for this book is available from the Library of Congress.

ISBN 0-7923-4840-0

Published by Kluwer Academic Publishers,
P.O. Box 17, 3300 AA Dordrecht, The Netherlands.

Sold and distributed in the U.S.A. and Canada
by Kluwer Academic Publishers,
101 Philip Drive, Norwell, MA 02061, U.S.A.

In all other countries, sold and distributed
by Kluwer Academic Publishers,
P.O. Box 322, 3300 AH Dordrecht, The Netherlands.

Printed on acid-free paper

Printed in the Netherlands

FOREWORD

This book is addressed to those who need to describe, compare, classify or analyse sets of texts, whether for research, study or teaching purposes. Texts can be literary texts, scientific texts (in bibliometrics, scientometrics, automatic information retrieval), economics texts, sociological texts (responses to open questions in socio-economic surveys), various types of interviews (in marketing, applied psychology, pedagogy, medicine), historical texts, political texts, etc.

We cover developments in *textual statistics*, a lively field of research whose exact boundaries are difficult to define, since the number of disciplines involved is so extensive, as is the assortment of potential applications. The chapters of this book attempt to emphasize this wealth of approaches, methods and fields, while presenting the state of the art of this field of endeavor.

Several ways of reading this book are possible, depending on the level of training of the reader, particularly the reader's knowledge of mathematics and statistics. A completely technical level is available to a person whose level is equivalent to a Masters in Economic Sciences. A practical user-oriented level is available to persons whose speciality is within the various potential fields of application.

The main text does not contain any strictly mathematical demonstrations. Instead, the interested reader is referred to more detailed explanations in specialized literature when available. On the other hand, definitions of concepts, implementations of procedures and rules for reading and interpreting results are explored in detail. The glossary of terms at the back of the book should help the reader understand the most important concepts and conventions for notation. Insofar as it was possible, we have abstracted the detailed mathematical proofs and placed them in the appendices.

This work is greatly indebted to various collaborations and work situations: in the Economics and Management Department of the Ecole Nationale

Supérieure des Télécommunications and in URA820 of the Centre National de la Recherche Scientifique (Traitement et Communication de l'Information); at the "Lexicométrie et textes politiques" Laboratory, URL 3 of the Institut National de la Langue Française (INaLF), at the Ecole Normale Supérieure de Fontenay-Saint-Cloud, and at the University Paris III (Sorbonne nouvelle).

We are grateful to, Hiroshi Akuto (University of Tokyo), Yasumasa Baba (Institute of Statistical Mathematics, Tokyo), Christian Baudelot (ENS, Paris), Monique Bécue, (Univ. Politecnica de Catalunya, Barcelona), Etienne Brunet (INaLF, Nice), Sergio Bolasco (Univ. La Sapienza, Roma), Conchita Callant (C.E., Brussels), Benoît Habert (ELI, ENS Fontenay/Saint-Cloud), Laurence Haeusler (Planistat, Paris), Chikio Hayashi (Institute of Statistical Mathematics, Tokyo), Georges Hébrail (EDF, Clamart), Nancy Ide (Vassar College), Dominique Labbé (CERAT, Grenoble), Max Reinert (Université Toulouse Le Mirail), Noboru Ohsumi (Institute of Statistical Mathematics, Tokyo), Marie Piron (ORSTOM, Bondy), Françoise Potier (INRETS, Arcueil), Jean Véronis (Université de Provence), for helpful suggestions and stimulating discussions on topics covered in this book.

Valuable comments on the manuscript were given by the reviewers. The present version of the text is widely indebted to their remarks and thorough correction of the text.

CONTENTS

INTRODUCTION

The methods for textual statistics brought together in this work have grown out of the meeting of several disciplines, principally: text analysis, linguistics, discourse analysis, statistics, computer science and socioeconomic survey analysis. Our endeavor derives from ongoing work in two areas: an activity with various names (lexical statistics, quantitative linguistics, etc.) that for the past fifty years has linked statistical methods to text analysis, and multivariate statistics, one of the main currents of modern statistics.

Researchers in a number of disciplines deal with large text sets requiring both text management and text analysis. Faced with a large amount of textual data collected in socioeconomic surveys, interviews, literary investigations, historical archives or documentary data bases, these researchers require assistance with organizing, describing and/or comparing texts.

We shall attempt to show how currently available computational and data management tools can be used to help describe, assimilate and finally evaluate data of a textual nature. The choice of a research strategy can only be made as a function of clearly defined objectives. What type of text are we analyzing? What questions are we attempting to answer? Is our goal to analyze a text's vocabulary in order to make stylistic comments about it? Are we looking for content through responses to a questionnaire? Are we attempting to determine motivations for purchasing a product through opinions expressed in focus groups or interviews? Or to classify documents in order to find them more readily?

Obviously, no single method can fully cover such diversified objectives simultaneously. However there is a series of methods that provide useful tools for achieving solutions to the above mentioned problems in a majority of textual analyses. In the present work, the purpose is to describe and explore textual data, and also to test hypotheses, uncover structural traits, and proceed to predictions. The field of application of the methods we present concerns processing responses to open-ended questions as well as textual corpora of a more general nature.

The availability of new fields of application may require new ways of preparing data for statistical processing, clearly defining concepts, properly managing algorithms and selecting specific results for presentation. This is particularly true as far as the domain of text analysis is concerned. In fact, in this domain, the concept of data, which is the basis for statistical counts, must be thought about in a very specific manner.

On the one hand it is necessary to cut up the units in a textual chain to carry out counts that can be used in subsequent statistical analyses. On the other hand, the textual chain cannot be reduced to a succession of disconnected units because much of the content of a text results from the relative positioning of words, and their possible juxtapositions or co-occurrences.

* *

The first chapter, *Textual statistics, scope and applications,* covers problems and approaches within the disciplines involved (linguistics, statistics and computer science). In each case it specifies the nature of the basic material constituting texts assembled into corpora.

The second chapter, *The Units of textual statistics,* is devoted to a discussion of the statistical units that lexicometric programs have to cut up or recognize (words, repeated segments). It introduces the properties of these units, which are the fundamental building blocks for the quantitative analysis of texts; it describes their respective relevancies as a function of the fields of application.

The third and fourth chapters: *Correspondence analysis of lexical tables,* and *Cluster analysis of words and texts,* present the basic techniques for exploratory statistical analysis of multivariate data with the help of examples that have been kept as simple as possible.

The fifth chapter, *Visualization of textual data,* applies the tools presented in chapters three and four to a description of associations among words and among demographic groups. It provides actual examples of applications, along with comments from a statistical viewpoint. It covers in detail the rules for reading and interpreting the results obtained, and points out their methodological range.

To complete these visual representations, the sixth chapter, *Characteristic textual units, modal responses and modal texts,* presents so-called characteristic words or characteristic element calculations that are used to identify the units whose frequencies are atypical in each of the parts of a corpus. An automatic selection of modal responses or modal texts makes it possible to restore words into their context, and to possibly characterize generally voluminous text parts with smaller pieces of text (sentences, paragraphs, documents, survey responses). Thus, in the case of free responses, the set of responses of a given group of respondents can be summarized through a few selected responses that have been identified as being representative of that group.

The seventh chapter, *Longitudinal partitions, textual time series,* covers the problem of *a priori* information concerning the parts of a corpus. In many applications, the analyst possesses external information that can be used to put together some of the parts, or to establish an order among the parts (textual time series). In this chapter the dependency relationships that can be observed among these structures and the lexical profiles of the parts are discussed through presentations of a method and examples of applications.

Finally, the eighth chapter, devoted to *Textual discriminant analysis,* studies the discriminatory power of texts in the statistical sense. How can a text be attributed to an author (or to a time period)? Is it possible to assign an individual to a group on the basis of his/her response to an open question? How can a document in a textual data base be classified (i.e., assigned to pre-existing categories)? In this chapter we attempt to show the contributions of textual statistics to stylometry, to information retrieval, and to some predictive models, through various examples.

The methodological journey on which we invite the reader is illustrated at each stage by textual corpora originating from different spheres of research. The results presented concern free response corpora in socioeconomic surveys and political speeches. The succession of examples are intended to allow the reader to appreciate the variety of actual and potential applications and the complementarity of the various processing methods. In addition, they should provide the reader with an opportunity to assess the applications of these methods in the literature.

1

TEXTUAL STATISTICS SCOPE AND APPLICATIONS

The study of texts using statistical methods constitutes a field of interest known as textual statistics. In recent years there have been important changes in the general context of this domain of research, as well as its objectives and the methodological principles it utilizes.

This chapter attempts to retrace the particular circumstances of the confrontation between linguistics and statistics, two disciplines that are fundamentally different in their principles and their past, both having undergone important changes, both deeply affected by the advent of the computer. It underlines certain aspects of the two disciplines that are likely to help understand their symbiotic relationship.

First we briefly review the distinct challenges facing linguists and statisticians, as well as experiences both disciplines have in common ; then we introduce the two fields of application with which this book is concerned: text analysis (literary, political, historical texts, etc.) and processing of textual corpora consisting of responses to open-ended questions in socioeconomic surveys.

1.1 APPROACHES TO PROCESSING TEXTS

We begin by situating textual statistics among the major disciplines that concern texts (*linguistics, discourse analysis, content analysis, information retrieval and artificial intelligence).* As we shall see, texts are an essential part of disciplines with different objectives, methods and research perspectives. A number of disciplines and fields of research (theory of languages, formal grammars, computational linguistics, etc.) use combinations of linguistics, mathematics and computer science without necessarily using the models and tools of statistics. They are mentioned here without much discussion.

1.1.1 The linguistic viewpoint

Linguistics, the "pilot science among the humanities" grew out of a schism with a whole series of prior practices in the field of the study of language. The concept of system plays a central role that virtually precludes examining isolated "facts". Structural linguistics involves describing linguistic elements within systems that assign different values to each of its elements. This "approach" is illustrated in the following excerpt from Ferdinand de Saussure[1].

> "Elsewhere there are things or objects that one is at liberty to consider from different points of view. Here the points of view come first, right or wrong, but only just points of view, with the help of which we create objects secondarily. These creations happen to correspond to reality when the starting point is right and do not when the starting point is wrong; but in neither case does any thing or object exist in its own right. Not even when considering the most material facts, the facts that are apparently the most obviously defined, such as a series of vocal sounds."

In the past century language was primarily analyzed through texts. *Philology* interprets and comments on texts by reconstituting the real meaning of the words of which they are composed [2].

After Saussure (1915) appeared, linguistics no longer considered text as its sole object of study. For a structural linguist, language itself is what matters: a set of systematic rules that cover combinations and substitutions of defined elements.

> When we separate language from speech we separate at the same time :
>
> 1) what is social from what is individual,
> 2) what is essential from what is ancillary and more or less accidental.[3]

Within linguistics there are several fields that study different aspects of language:

---

[1] Notes dating from 1910 that appeared in the *Cahiers Ferdinand de Saussure*, 12 (1954) quoted by Benveniste (1966).

[2] As Pêcheux wrote (1969): One wonders simultaneously, "What is this text about?", "What are the main ideas contained in this text?", and, at the same time, "Does this text conform to the norms of the language in which it is presented?", or, "What norms apply to this text?" /.../ In other words, the classical science of language claimed to be at the same time the *"science of expression"* and the *"science for the means of this expression"*.

[3] Cf. Saussure (1915).

- *phonetics* studies the sounds of language, whereas *phonology* studies phonemes, that is, sounds as distinctive units.
- *lexicology* studies words, their origins, history and inter-relationships.
- *morphology* studies words independently from their associations within sentences. It studies morphemes or variable elements in words: grammatical morphemes (word endings or inflections) and lexical morphemes.
- *syntax* studies relationships between words in sentences (order of words, agreement).
- *semantics* studies the meaning, the content of the message.
- *pragmatics* studies associations between utterances and the circumstances of communicating.

Note that in practice the rules defined within each of these fields are connected not only among themselves but with the rules of other fields. Thus lexical studies cannot be made without reference to the meaning of words (semantics). The study of word meanings includes their function in sentences (syntax)[1] and sometimes the study of the context of expression (pragmatics).

1.1.2 Content analysis

Several approaches to text analysis developed after the study of the meaning of texts was abandoned by linguists. *Content analysis,* developed in the United States at the beginning of this century is applied primarily to studies related to press documents. According to Berelson and Lazarsfeld (1948) it is considered:

> "/.../ a research technique for the systematic and quantitative objective description of the apparent content of a communication."

Content analysis (see, for example, Weber, 1985) proposes to directly access the meanings of different segments which make up a text, without considering the textual material itself. Thus it operates in two phases. In the first phase the coder defines a set of equivalent categories or themes, the occurrences of which are then counted in the text being analyzed. In the second phase the coder performs counts on each of the themes identified in the initial grid. Elements counted by content analysis can be themes, words, or even syntactic or semantic traits. The counting unit for quantitative measures also varies: word, physical area covered by article, etc.

As we can see, content analysis thus defined involves a statistical dimension. However its successful use supposes that the system of categories defined *a priori* is both coherent and relevant, both of which are difficult to ensure in practice.

---

1 Sometimes the phenomena that have to do with morphology and syntax are grouped together under the name of morpho-syntactics.

1.1.3 Artificial intelligence

Research programs in *spoken and written word comprehension* have far more ambitious objectives. This extensive field of investigation is at the heart of activities designated by the expression *natural language processing*. It involves a number of endeavors of obvious economic value such as machine translation, spelling checkers, speech and handwriting recognition, natural language user-machine interfaces etc.[1]

Although these fields of analysis sometimes cross over into one another, (for example, automatic information retrieval and bibliometrics use statistical tools as well as procedures drawn from artificial intelligence), the concerns are nevertheless distinct and the approaches complementary.

In the field of artificial intelligence, texts are analyzed for "real-time" or "decisional" applications (the term "natural language" is used) so as to facilitate or improve the man-machine dialogue. For computer scientists working in this field, the goal is to obtain a representation of the meaning of sentences that are presented to the computer system.[2]

The concept of comprehension, which occupies a central position in most research involving artificial intelligence and texts, is virtually foreign to textual statistics, except in some cases for pre-processing phases which are discussed in chapter 2. On the other hand, the independence of the methods of textual statistics from the phase of "understanding" texts is illustrated in the course of the following chapters through application examples that use the same software tools to deal with English, French and Japanese. Of course this independence is only temporary: results achieved through counts and automatic statistical processing become additional material to be added to the overall analysis material.

1.2 STATISTICS AND TEXTS

Successful applications of statistical methods in numerous fields of the natural sciences (physics, biology, etc.), in the humanities (psychology, economics, etc.) and in disciplines that deal with the use of language, gradually came to the attention of specialists in the field of vocabulary studies during the nineteen fifties and sixties.[3]

---

1 Cf. Partridge (1991). Carré *et al.* (1991) gives an overall view of the problems and results in these effervescent fields of research.

2 A summary of *Natural Language Discourse Processing* is published by McKevitt et al. (1992).

3 It even seems as though the lack of statistical analyses was perceived at first as a "delay" with respect to other disciplines that had to be caught up as quickly as possible. "It seems to me that it would be a barrier to the development of linguistics if we continue to be so disinterested in numbers when we speak of linguistic phenomena"; (Cohen, 1950).

Initially, it was discovered that lexical distributions followed empirical rules that could be used to improve stenographic transcriptions (Estoup, 1916). Later they were studied under the title of "psychobiology of language" by Zipf (1935).

1.2.1 Pioneering works

Analyses that used statistical methods for text analysis often applied a series of quantitative approaches to the linguistic units (phonemes, words) counted in a text.

Lexical statistics (Yule, 1944; Muller, 1964) attempted to resolve a series of problems raised in stylistic analysis (comparative studies of the vocabulary of the "great authors").

Methods developed in the same framework, and presented by G. Herdan under the name of *statistical linguistics,* claimed to be "a quantification of de Saussure's theory of language". According to Herdan (1964) this discipline is a branch of structural linguistics and its main function is the statistical description of how elements defined by linguists at different levels of linguistic analysis (such as the phonological, lexical, or sentence levels) function within textual corpora.

This simplification, advanced by Herdan as a theory, appears to be somewhat open to discussion when it is compared to the approach of the founder of structural linguistics as set forth above.

1.2.2 Information retrieval

Information retrieval constitutes its own specific field of research that is at the cross-roads of computer science (essentially: data base research), artificial intelligence, quantitative linguistics and statistics.

Statistical methods may be called upon at the time the data base is built and organized (aids to syntactic analysis and categorization)[1], in exploratory and *navigational* phases within the data base[2] and finally in the information retrieval phase based on descriptive elements in natural language or on key-words.

This last phase is covered in more detail in chapter 8, concerning methods of textual discriminant analysis. *Bibliometrics* and the *technological watch* are two disciplines that encompass information retrieval as a basic tool.

---

1 In addition to the basic work of Van Rijsbergen (1980), Salton and MacGill (1983), see, for example, Blosseville et al. (1992), Lewis et al. (1990,1992), Wilks et al. (1991).

2 Cf., for example, Cutting et al. (1992).

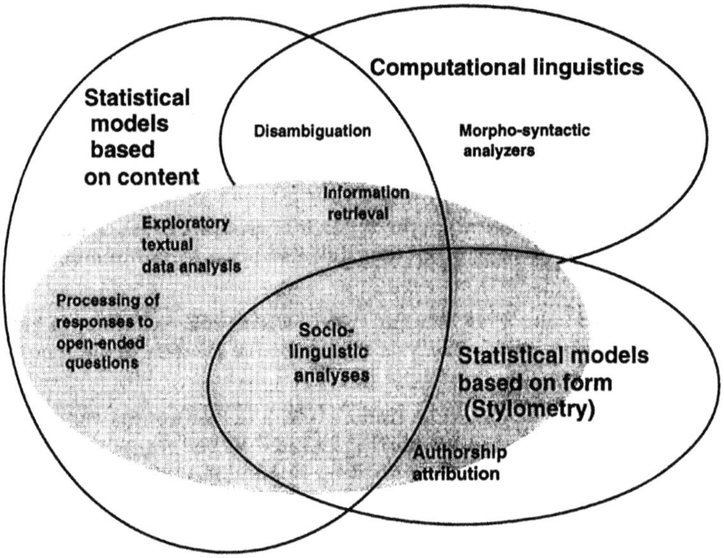

Figure 1.1
Scope of exploring textual data (outlined by gray area)

1.3 STATISTICAL PROCESSING OF TEXTS: A STATISTICIAN'S VIEWPOINT

As a first approach to the complex relationship between statistics and texts, we begin with a few concepts from the statistician's viewpoint. According to statisticians, texts should be analysed through the use of discrete, qualitative variables such as counts rather than with variables of a continuous nature.

However, methods for analyzing qualitative variables are generally more complex and more difficult to implement than methods for analyzing continuous variables. These methods for analyzing qualitative variables were developed extensively during the nineteen sixties, stimulated by the new computational capabilities afforded by the computer. These circumstances might explain the recent upsurge of applications of multivariate qualitative statistical applications to the field of text analysis [1].

1.3.1 Processing phases

In most statistical applications a sequence of four steps is generally executed, symbolized as follows:

problem - data - processing - interpretation

---

[1] Cf. Benzécri (1973, 1981).

Each phase of this sequence raises issues that vary with the context, problems and fields of application.

The *problem* that motivated the study can give rise to an *a priori* formalization of a statistical or probabilistic model, or on the contrary, it can be stated in very general terms, dealing initially with a summary description or exploration of the universe under consideration.

The *data* can be experimental (in fact data collection can be a function of the model, for example in the case of experimental design); or they can arise from observation (in some cases they may already exist before the formulation of the problem, as in the case of analyzing secondary data in surveys, for example).

The phase designated as *processing* comprises, in the most classical inferential case, testing hypotheses or models. In the case of descriptive or exploratory analysis, this phase is primarily a rearrangement of the data in order to uncover its principal structural traits.

Finally the *interpretation* phase can be reduced to taking into account the conclusions of a hypothesis test, in the most classical case; often it involves a critical evaluation of hypotheses and of the initial model, if one exists. In the case of descriptive or exploratory analysis, the interpretation phase inevitably includes a consideration of the validity and significance of observed structures, with sometimes an evaluation of the data (relevance, collection, coding) and a rethinking or refining of general hypotheses. These evaluations and considerations may give rise to new treatments, new observations, possibly new models. Obviously, this description only sketches the statistician's activities in its broadest strokes. In practice there are numerous other situations. In fact these situations are the ones that complicate the user's task, and make the work particularly challenging for researchers.

1.3.2 Internal and external information, meta-data

The need to automate the *problem-data-processing-interpretation* sequence in view of establishing a statistical phase in expert systems has recently led some researchers to introduce the concept of *meta-data* or *meta-information.*

Briefly, there exists a large amount of information concerning a data matrix about to be analyzed that does not appear in the matrix itself. Thus, inequalities, relationships between values of certain variables, or boundaries (upper and lower) for these values are the most elementary forms of meta-information. In processing a survey, this meta-information, which is relatively easy to formalize, is used routinely to check and "clean" data files or to carry out consistency tests[1]. The development of exploratory

1 Attempts to formalize meta-information were carried out by Hand (1992), and in the context of information retrieval, Froeschl (1992).

analyses and work done on data bases has accentuated interest in the concept of meta-information. The rediscovery of known structures is useful for verification purposes, but does not constitute the ultimate goal of these analyses, which have to learn to use what is already known in order to gain more knowledge.

1.3.3 A wealth of meta-information

This all too brief description is meant to situate textual information within the normal context of statistical activity: meta-information is particularly abundant in the case of textual data. Every word, even a function word is allocated several rows or several pages in an encyclopedic dictionary.

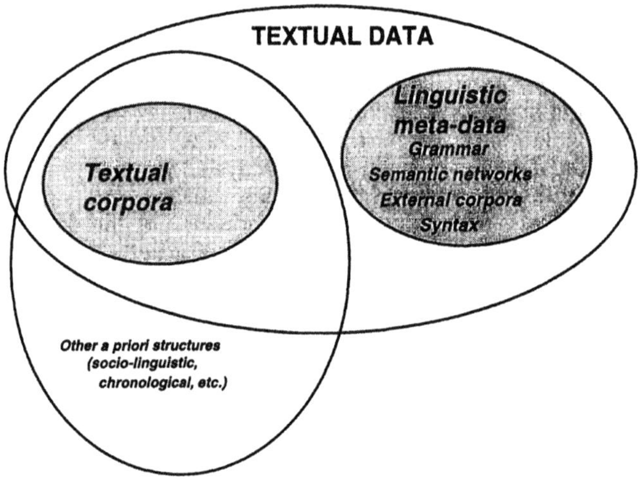

Figure 1.2

Textual data and external information

This concept of a *function word* (sometimes referred to as empty word, or tool word or grammatical word in information retrieval) is widely used by text researchers. Most of the time, discarding function words is a practical way of limiting investigations to words with a greater semantic content. It is however impossible to draw up a definitive list of this type of word that would satisfy the requirements of all textual studies[1].

---

[1] Note the clearcut distinction (also stressed below in chapter 8) between the concerns of stylometry (often emphasizing the role of function words as author trademarks, for example) and content oriented statistical analyses, that tend to minimize the importance of function words as relevant statistical units.

The rules of grammar obviously constitute basic meta-information (figure 1.2). Words belong to semantic networks that dictionaries and partially automated morpho-syntactic analyzers attempt to take into account. The main issue is whether these different levels of meta-information are relevant to the problem under consideration.

Suppose, using an example outside the textual field, that the histograms of wavelengths of the colors of a Rembrandt painting are analyzed (using each pixel of a reproduction). It goes without saying that only a minute fraction of the information contained in the original image is utilized. However it is possible that the shape of the histogram (or of a more elaborate function of the same measures and basic data) would suffice to distinguish between a Rembrandt and a Rubens or a Van Dyck. The entire set of available meta-information is not indispensable and therefore does not need to be used, if the only objective is to discriminate between paintings.

Similarly, stylometric studies discussed chapter 8 (Shakespeare's poems) only use an infinitesimal part of the information contained in the texts concerned, and an even smaller part of the knowledge we have about these texts.

On the other hand, in some information retrieval applications, for example, it is possible to work with key words that are treated as classical presence-absence qualitative variables, and thus to construct tables that are wholly analogous to those encountered in other statistical applications. Then the document to be categorized or retrieved is no longer a text, but a bundle of words, without order or syntax.

Modelling frequency distributions

Even if the above-mentioned meta-information is not taken into account, textual material constitutes a relatively complex entity for statisticians. Every text possesses a sequential, or syntagmatic dimension, which relates it to what is designated in statistics under the name of process[1], although the formulation of a text constitutes a particularly complex form of process. In addition, the words of a text have relationships among themselves (and also with other words of the same language) that can be brought to light through counts performed on the entire text.

Several models for the theoretical distribution of vocabulary have been proposed since the inception of statistical analyses applied to texts. The earliest attempts were made by Zipf in 1932 (cf. chapter 2, section 2.3.2) and Yule (1944). Waring-Herdan's distribution (cf. Herdan, 1964; Muller, 1977) is probably the most often quoted model in the linguistic literature. A more recent example of a model for frequency distributions of vocabulary is given in chapter 8 (section 8.3) concerning stylometric studies.

---

1 In chapter 2 we shall see how the use of statistical elements called repeated segments makes it possible to account operationally for some sequential aspects of a text.

Given the complexity and the very nature of the basic material, these models only serve as formal analogies, and do not in any way claim to participate in explaining the phenomenon of text creation. However the exploratory methods that provide the most heavily used tools of textual statistics, and that occupy a central position in this book, do not rely directly on these methods.

1.4 SPECIAL TEXTS: RESPONSES TO OPEN QUESTIONS

Open-ended questions can be found in the questionnaires of a multitude of surveys performed in socioeconomy, epidemiology, advertising, business or political marketing. They become an essential part of these questionnaires when the scope of research goes beyond a simple tally, and when a complex and unknown topic is being explored.

Processing free responses (i.e., responses to open questions) is a challenge for both statisticians and specialists in text analysis. In a corpus of free responses, lexical frequencies are artificial for the most part, because the same question is asked among hundreds or thousands of people (cf. figure 1.3). The juxtaposition of the responses results in a redundant text by construction, in which stereotypes are not uncommon.

In what follows we discuss the advantages and disadvantages of this type of information. However our main thrust is to elaborate on their statistical processing.

Lexical frequencies in free responses

A thousand responses to the question: "*Do you watch television every day?*" constitutes a text where the forms *yes* and *no* are predominant, and where the relative frequencies of these forms has a simple interpretation that is familiar to survey specialists.

Responses to an ancillary question "*Why?*", asked after the preceding question have an intermediate status. Since they constitute one thousand identical stimuli, these responses can be stereotypical or they can include original and unexpected contents and expressions. Even if expected form differences are taken into account, simple counts are notoriously inadequate. On the other hand, when responses are grouped within categories (for example, age, gender, profession) comparisons of the mean lexical profiles of these categories can be revealing.

What can be done with textual information consisting of one or two thousand free responses (entered into the computer as is, without transformations or coding)?

The most common approach consists in transforming the responses by "closing" the open question *a posteriori*. This practice is called *post-coding*, a manual technique whose main characteristics we examine briefly below. This technique unfortunately contributes to maintaining the dangerous illusion that there is a similarity between closed-end questions asked during the interview and questions that have been closed for coding purposes, whereas in fact these two types of questions are not really comparable.

A methodology concerning the use of *open* questions is briefly discussed in section 1.4.1 below.

After setting aside the idea of manual (subjective) coding performed before textual data is entered into the computer, we shall define the various types of statistical units used for coding and analyzing textual information. This point is developed in a more general framework in chapter 2, which is devoted to the units of textual statistics.

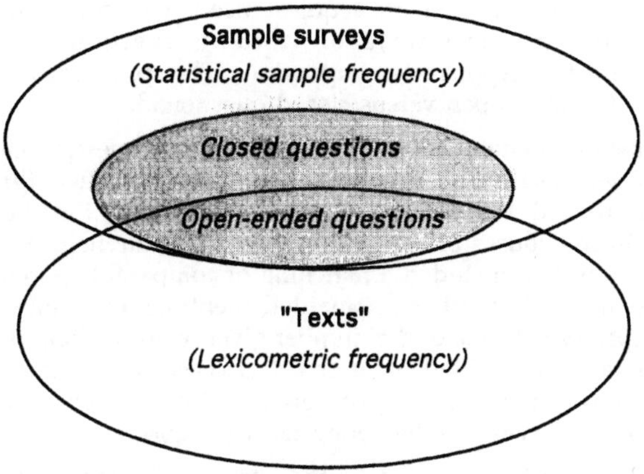

Figure 1.3

Status of frequency in the case of open-ended questions

1.4.1 Open-ended questions: a research tool

In order to thoroughly evaluate the informational advantages offered by free responses, we first compare the two types of questions, open and closed, and we show that they produce information of a different nature that is not readily comparable. Then we examine cases in which open-ended questioning is inevitable, either for technical reasons, or because of the very nature of the information sought. After briefly covering post-coding techniques, we discuss procedures aimed at forming corpora out of various aggregates of free responses.

Open versus closed questions: a brief reminder

Numerous works concern comparisons between open and closed questions from the point of view of their relevance and their efficiency: most of the findings are essentially empirical, and they generally uncover important variations having to do with the actual content of the questions.

It is known that the *wording* of a question plays a fundamental role in an attitude or opinion questionnaire. In fact, it would be difficult to find a case where two distinct ways of wording the same closed-end question produce the same results in terms of response percentages.

The sensitivity of response percentages relative to wording is particularly strong in the case of opinion questions. Thus the classical experiment of Rugg (1941) states that the response *"yes"* to the question: *"Do you think the United States should forbid public speeches against democracy?"* receives 21 percentage points less than the response *"no"* to the question: *"Do you think the United States should allow public speeches against democracy?"*[1].

Because of this, analyzing systematic acquiescence is particularly difficult. It seems that finding practical ways of wording questions to produce equivalent *yes* and *no* responses is impossible. This problem of different wordings is also found in open versus closed questions [2].

However, the lists of items attached to a closed end question can play a positive role if recollection is involved: this was established through an experiment performed by Belson *et al.* (1962) concerning newspaper headlines read in the course of preceding days. With the help of an actual experimental design, that included a matching of comparable populations, it was established that the number of unaided mentions of dailies read the previous day represented *71%* of the number given with the help of a check-list. This percentage of unaided mentions decreases to *25%* in the case of weeklies and monthlies, which somewhat disqualifies open-ended questioning when a survey involves long-term memory.

Conversely, in very particular contexts, the absence of a response item can also play a positive role. It can establish a climate of confidence and communication, and lead to better results when certain subjects are brought up.

This is what is indicated by the work of Sudman and Bradburn (1974) concerning questions having to do with "threats", and of Bradburn *et al.* (1979) concerning questions about alcohol and sexuality.

---

1 This absence of symmetry between the two wordings, which was verified on other themes, is stronger among persons whose educational level is lower.

2 A classical experiment, quoted by Schuman and Presser (1981), demonstrates this in a convincing manner. When asked *"what is the most important problem facing this country [USA] at present"*, *16%* of Americans mention *crime* and *violence* (grouped free responses), whereas the same item asked in a closed question produces 35% of the same response. The explanation given by authors is the following: lack of security is often considered as a local, not a national problem, so that the item *crime and violence* is not mentioned spontaneously very often. Closing the question indicates that this response is a relevant or *possible* response, resulting in a higher response percentage.

When should open questions be used?

Sociologists like Lazarsfeld (1944) suggest the use of open questions primarily in the preparatory phases of a study, to develop a battery of response items for a closed-end question. This utilization is still recommended, but rarely implemented because of its cost: obtaining a list of items that include those that are relatively infrequent requires a rather extensive pilot study, which financial constraints and the need for quick results often do not allow.

There are three typical situations for which open questions must be used:

- *To shorten interview time:* Open ended questions are less costly in terms of interview time, and generate less fatigue and tension. In the case of voluminous lists of items, the duration of the interview can be sharply reduced.

- *To gather spontaneous information:* Marketing survey questions contain many questions of this type. For example: " *What do you recall about this ad? What are your feelings about this car?"*.

- *To probe the response to a closed-end question.:* This is the follow up additional question *"Why?"*. Explanations concerning a response already given have to be provided in a spontaneous fashion. A battery of items might suggest new ideas that would only mar the authenticity of the explanation given. The importance of this additional information has been advocated by many sociologists who specialize in surveys such as Schuman (1966), who alludes to international surveys where problems of comparability and wording comprehension are acutely present. In international studies, it is important to know whether people interviewed in different countries understand the questions in the same way. In fact it is legitimate to raise this same issue with respect to regional, generational and socio-cultural differences.

1.4.2 Manual *post-coding* of free responses

This widespread technique for processing free responses consists in constructing a battery of items on the basis of a sample of responses (generally *100* to *200*), and then coding the whole set of responses, replacing each open question with one or several closed questions. This procedure has a significant advantage: the results are readily processed.

Let us however mention a few drawbacks of this type of processing.

Coder bias: Coder bias is added to interviewer bias, since the coder makes decisions and formulates interpretations, introducing a "personal touch": decisions are sometimes difficult and controversial from the specialist's viewpoint.

Alteration of form: Information is destroyed in its form and often weakened in its content: quality of expression, level of vocabulary, and general interview tonality are lost in post-coding.

Weakening of content: When questioning allows for responses that are composed, complex, vague and diversified, information is literally destroyed by post-coding; and yet this is the case where the heuristic value of free responses is the greatest. Responses should be allowed to retain their form, their texture and their tonality more often than is commonly thought, in order to really be understood by the researcher. In any case, complex responses, even responses formed by juxtapositions of elements that are easily coded, are very difficult to analyze[1].

Infrequent responses are eliminated a priori: Infrequent, unique and unclear responses are assigned to "residual" items which are extremely heterogeneous and thus lose all operational value. This criticism of post-coding is concerned with its preliminary nature. Post-coding occurs at the outset, before analyses are carried out, so that many decisions of assignment and grouping are made (often by non-specialists) without analyzing the entire set of material collected in all of its complexity and richness [2].

1.4.3 Artificial texts: groups of responses

The type of treatment proposed to remedy the weaknesses of post-coding does not involve an *a priori* reduction of raw information, but on the contrary imparts added value to this information by using all other available information about respondents (which information is considerable in the case of a survey) and all the computer's options for data management, sorting and computing.

As free responses are entered into the computer in their original form, they can be matched with interviewees' demographic characteristics as well as their responses to closed questions. Then they can undergo data management procedures that are both useful and elementary, such as categorizations and groupings, without being altered.

For example, responses can be grouped by socio-professional categories. Thus responses given by farmers, entrepreneurs, housewives, workers and

---

1 For example, take a response to a "Why?" after a rather general question about hopes for future years: *"That we should be in good health and that our children should be happy, this is what is most important to us; everything else doesn't matter much"*. The items *health* and *children's happiness* can be identified and recorded. But how does one code the idea of excluding everything else, which is a fundamental, characteristic concept in the content of this response in terms of life style?

2 Thus, in response to a question concerning *"hopes for future years"*, an infrequent item such as *fewer injustices* is not kept at first. In fact, it is mentioned with significant frequency (that is, with a frequency that cannot be imputed to chance or to sampling fluctuations) by farmers, who are a minority in the sample. Therefore these infrequent items merit being coded.

upper management can be examined separately. There might be categories or combinations of categories that are relevant with respect to each open-ended question. By grouping responses within categories, "artificial speeches" are obtained that are all the more meaningful when the categories are carefully chosen. Reading and interpretation are made easier in that repetitions and concentrations of certain issues appear within each category. However this rearrangement of raw information can be carried out in many different ways. The problems are, first, deciding how to group responses in a relevant manner, and second, facilitating the interpretation of the groupings thus generated.

How to group responses?

Obviously the importance of the mode of grouping is considerable: there are several possible strategies for finding one or more relevant partitions. These strategies are complementary and gain from being used simultaneously.

First one can use the criteria that are thought to be the most discriminating through prior knowledge of the theme being analyzed, with or without the use of cross-tabulations. If, for example, the questions relate to the evolution of the family, and if an age or generational effect combined with a sociocultural effect is suspected, a compound variable that cross-tabulates age and educational level can be developed.

Also a partition can be sought that is as universal as possible, within the limitations of sample size: this is the principle behind *working demographic partitions* discussed in sections 4.3 and 5.2. Major characteristics that are judged to be relevant (for example: age, category, gender, educational level, region) are combined through an automatic clustering technique in a single partition: this is equivalent to reducing several thousand individuals to approximately twenty to fifty groups that are as homogeneous as possible with respect to the above-mentioned criteria.

On the other hand a direct typology of responses can be carried out (without preliminary grouping) on the basis of their lexical profiles (this only makes sense if the responses include more than two or three words). Then categories that have the highest association with this typology can be selected before proceeding to groupings. These different strategies are discussed in detail in chapter 5.

Analyses of *grouped responses* are rather similar to analyses of literary, political and historical texts, whereas analyses of *individual responses* are similar to treatments carried out in information retrieval. The originality of the approach lies in the large number of grouping possibilities and thus the large number of viewpoints accessible to the researcher.

The elementary operation of aggregating responses greatly facilitates reading the original text. However, reading hundreds or thousands of responses for each partitioning of respondents remains an expensive task when routine analysis rather than in-depth research is involved. In this case it is helpful to have a series of tools for comparing bodies of text obtained through groupings. The analyst most certainly wishes to know the characteristic words for a given category; he/she also wishes to know which groups express themselves in similar ways.

To satisfy these needs, the textual material has to be prepared and dissected in such a way as to define new elements that are likely to be recognized and treated by computational programs.

2

THE UNITS OF TEXTUAL STATISTICS

The need to compare texts quantitatively arises in many different disciplines of scientific research. In each particular case, the use of quantitative methods is motivated by different requirements and often very divergent objectives (comparative stylometric analyses of texts written by different authors, typologies of individual responses to a single open question, information retrieval, etc.).

However, experience with applying lexicometric processing methods to texts originating from different sources has shown that a single series of methods has relevance in many diverse text-based studies. The following chapters are dedicated to presenting these methods.

2.1 CHOOSING UNITS

Segmentation, identification, lemmatization, disambiguation

Statistical methods rely on measurements and counts based on objects that are to be compared. Counting elements and adding them together means that they are being treated as either identical occurrences of a the same type or identical occurrences of a more general *form*, at least for the duration of the experiment.

In practice, applying these general principles requires defining a *norm*, whose purpose it is to isolate the various units within the chain of text upon which counts are carried out. The operation of subdividing the text into *minimal units* (that is, units that are not to be subdivided further) is called *segmentation* of the text. This phase of breaking the text up into distinct units is followed by a phase of *identification*, that is, grouping identical units together.

Different counts have different degrees of relevance in each particular field of research. They also have different advantages (and disadvantages) as far as practical implementation is concerned.

For instance, a researcher exploring a set of articles assembled from a data base in the field of Chemistry might require that the noun *acid* and its plural *acids* be grouped into the same unit, in order to be able to query all of the texts at once on the presence or absence of one or other word.

On the other hand, in the field of political text analysis, researchers have observed that singular and plural forms of a noun are often related to different, sometimes opposite, concepts (cf. for example the opposition in recent text of *defending freedom/defending freedoms*, referring to distinct political currents). In this latter case, it could be often preferable to code the two types of elements separately and to include both in the analysis.

Beyond these considerations of fields of research, once the segmentation norm is established, *textual statistics* methods are applied universally to counts developed on the basis of any norm. However, the textual units' origins should remain present in the researcher's mind during subsequent phases of analysis.

2.1.1 Computerized text

Every computer used for loading and storing text has a system of characters consisting of about one hundred elements. Some of these characters are letters of the alphabet: uppercase and lowercase letters, language-specific letters with diacritics (accents, for example); others are codes for numbers; others are codes for signs such as percentage or dollar sign; finally there are punctuation marks. The "carriage-return" character is used for separating *paragraphs*.

In practice, a set of typographical norms is gradually being established for performing the operation that was previously referred to as *encoding* texts. These norms are increasingly influenced by technological progress in the field of text computerization. Thanks to scanners, the volume of texts within the reach of researchers for lexicometric analyses cannot be compared to what was available to them just twenty years ago. Because of these circumstances it is essential to have at hand relatively simple methodological tools for comparing and analyzing ever expanding textual information.

2.1.2 Analyses based on graphical forms (or words)

A particularly simple way of defining textual units in a corpus of texts is to analyze graphical forms (or words, or types). This approach can be used for various purposes depending on the objectives of the analysis: verification of data entry, inspection of vocabulary, creation of a data base for subsequent statistical comparisons. To obtain an automatic segmentation of a text into occurrences of graphical forms, a subset of characters must be specified as *delimiting characters*. (all other characters contained in the text are then considered as non-delimiting characters).

A series of non-delimiters whose bounds at both ends are delimiters is an *occurrence*. Two identical series of non-delimiters constitute two occurrences (*tokens*) of the same word (*type*). The entire set of words in a text is its *vocabulary*. In the context of this type of segmentation, a text is a series of occurrences that are separated from one another by several delimiters. The total number of occurrences contained in a text is its *size* or its *length*. This approach assumes that each character in a text has only one status, that is, that the text has been cleaned of any coding ambiguities (for example: periods at the end of a sentence and points occurring within abbreviations: U.N.E.S.C.O., I.B.M., etc.).[1]

2.1.3 Lemmatized analyses

In the lexicometric approach, the graphical elements resulting from automatic segmentation may be *lemmatized*. This means that identification rules must be established in order to group together words arising from the different inflections of one lemma.

The main steps for lemmatizing the vocabulary of a text written in English are as follows :

- verb forms are put into the infinitive
- nouns are put into the singular
- elisions are removed

However, the actual lemmatization of the vocabulary of a corpus involves some unavoidable problems that may be difficult to resolve[2]. It is relatively easy to reconstruct the infinitive of a verb from its conjugated form, and a singular noun from a plural noun. However, a systematic determination of the lemma to which each graphical form belongs in a text requires prior disambiguation in many cases.

Ambiguities arise when a given string corresponds to inflectional forms of different lemmas (for example: *can* from the verb *can*, and *can*, singular noun). In some cases several derivations may exist from the same etymological source (cf. the different meanings of the word *state*, for example). In other cases, ambiguities concerning the syntactic function of a word have to be removed, requiring a grammatical analysis of the sentence containing it. Some ambiguities of a semantic nature can be removed through simple inspection of the immediate context. Others require examining several paragraphs, or even the text in its entirety. Sometimes

---

1 Attention should also be paid to differences introduced by the use of uppercase letters (beginning of sentence or proper name), by hyphens, etc.

2 An automatic lemmatization requires a disambiguation, involving a morpho-syntactic analysis and sometimes a pragmatic analysis. See Charniak (1993), Gale et al. (1992), Weischedel et al. (1993), Bolasco (1992).

ambiguities can exist between several meanings of a word. They might even be intentional, in which case an arbitrary choice has to be made.

Muller (1968, p. 143) discusses the difficulties associated with establishing this type of norm for analysis.

> The norm should be acceptable at the same time to linguists and their adjuncts as well as to statisticians. However their needs are often contradictory. Linguistic analysis results in loose classifications that always include indeterminate zones; the material it uses is eminently continuous, and only rarely is it possible to find clear boundaries; most of the time it requires a close inspection of the syntagmatic (contextual) and paradigmatic (lexical) surroundings before subdividing. On the other hand, in all its applications, statistics becomes involved in simplifying categories; it can only do its work once the continuity of language has been made discontinuous, which is harder at the lexical level than at other levels; it loses its effectiveness when distinctions increase or categories become smaller; when working on large data sets it has difficulty dealing with the type of thinking that prefers analyzing isolated facts.

An examination of the problems associated with lemmatization shows that it is impossible for a method to be fully effective as well as totally automatic in generating lemmas for each unit obtained by segmenting a text into graphical forms.

2.1.4 Semantically based approaches

This heading brings together various methods used in often unrelated fields, that have in common their substantial pre-coding of initial texts. The goal of this operation is to *tag* a text with a series of semantic traits associated with each of the elements that compose it, using procedures defined in a rather formal way.

Frames

Frames, proposed by Roger Schank (1975), are meant to "represent the meaning of each sentence". This operation is an extreme example of a methodology that tends to insert speech into semantic categories that are defined *a priori*. Schank's representation is supposed to anticipate, at least in a general way, all the questions that might be asked about a textual unit in the framework of what he calls *everyday life*. Various publications in the field of Artificial Intelligence bear witness to the utility of such undertakings for setting up expert systems in particular fields of knowledge.

However our understanding of the problems and practices concerning text analysis in many fields of social science leads us to believe that such a pre-analysis categorization may result in a mutilation of the textual material.

Content analysis

Although *content analysis* uses a more complex coding scheme of the real world, it is not too dissimilar from what we have discussed above. In this case, which we already briefly covered in chapter 1, textual units are grouped into *a priori* groups before carrying out counts or after an initial reading of the text, so that counts are directly related to "content". Problems may arise because of the unlimited latitude that is given the user in defining categories for counting.

2.1.5 Brief comparison with other languages

Problems associated with using graphical forms in lexicometric analyses become clearer when various languages are compared. The examples that follow are intended to introduce the reader to the variety of problems that arise when the counting system uses automatic delimiters. As we shall see, differences between languages include both morpho-syntactic particularities that are intimately involved with the structure of each language and the orthographic tradition that belongs to each language.

For example, in contrast to French, English uses a single graphical form to represent many inflections of a verb form: (*speak* versus: *parle, parles, parlons, parlez, parlent,* etc.*)*. The same is true of inflections representing the gender of an adjective in French.

In Spanish, personal pronouns: *nos, se,* enclitic pronouns, are appended to the end of the verb forms with which they are used: (*dámelo* versus *give it to me*). This mechanism, which gives rise to a large number of different graphical forms, also complicates the task of finding personal pronouns as entities in a text.

Languages with declensions allocate the different cases of a given noun into a number of different graphical forms. Thus, Russian scatters the inflections of a noun into several graphical forms according to its grammatical function in the sentence, thus economizing a great deal on the use of prepositions.

министры	ministers
совет министров	counsel of ministers
председатель совета министров	president of the counsel of ministers
председателя совета министров[1]	of the president of the counsel of ministers

Depending upon one's interest in identifying grammatical categories in a text for lexicometric analysis, this circumstance can be considered fortunate or, on the contrary, a major impediment to detecting the various inflections of a textual element.

In addition to declensions, German has the feature of creating compound words by concatenating several nouns, or the root of a verb and a noun.

Trinkwasser	drinking water
Zusammengehörigkeitsgefühl	feeling of belonging together

---

1 Transliteration: ministri, soviet ministrov, predsedatel sovieta ministrov, predsedatelia sovieta ministrov.

In the following list of compound nouns in a text containing the form *Bildung (formation)*, the question as to whether to break up each of these compound elements into more elementary units before performing lexicometric analyses on texts is not straightforward.

Bildung	formation, form, culture
Hochschulbildung	college-level teaching
Weiterbildung	continuing education
Preisbildung	determination of prices
Ausbildung	culture
Wordbildung	morphology (lexical)
Gesichtsbildung	shape of face
Herzensbildung	nobility of soul

To complicate matters further, the same text may contain compound words for which the common form is placed at the beginning of the word. Experiments we have conducted on texts in different languages have shown us that most of the time a language's morpho-syntactic idiosyncrasies are not a major obstacle to analyzing texts with the methods of textual statistics. As we shall see further, typologies resulting from textual counts are relatively insensitive to variations in the counting unit.

In conclusion, the graphical form is in no way a *natural* element for analyzing texts; the advantage of counts based on graphical forms is that it is remarkably easy to automate the process.

2.2 SEGMENTATION AND NUMERIC CODING OF TEXT

Computer-based processing of textual data is greatly simplified by applying a basic technique called: *numeric coding*. This technique consists in giving a numeric code to each word to be used during calculations. This code is associated with each occurrence of the word. Code are stored in a dictionary of words that is unique for each application.

The Life corpus

The *Life* corpus consists of a set of responses given by *1000* English-speaking respondents[1]. The wording of the question is: *"What is the single most important thing in life for you?"*, followed by the probe, *"What other things are very important to you?"*.

Table 2.1 shows examples of responses to this question along with their associated codes for three demographic variables. The numeric coding of the text is based on formal criteria such as: alphabetical order, number of characters in a word, etc. In what follows we use a combination of criteria to assign numbers to the text's words.

---

1 For more information about this survey, cf. Chapter 5, section 5.1.4.

Table 2.1

Examples of responses to Life question

Gender	*Educ.*	*Age*	***Responses***
1	1	4	happiness in people around me, contented family, would make me happy
1	2	2	my own time, not dictated by other people
1	2	2	freedom of choice as to what I do in my leisure time
1	3	2	I suppose work
1	2	1	firm, my work, which is my dad's firm
2	1	6	just the memory of my last husband
2	2	6	well-being of my handicapped son
1	1	5	my wife, she gave me courage to carry on even in the bad times
2	2	3	my sons, my kids are very important to me, being on my own, I am responsible for their education
1	3	3	job, being a teacher I love my job, for the well-being of the children
1	2	3	job, sometimes it is very hard to find a job
2	2	2	peace of mind
2	1	4	good health
2	1	2	absence of money worries
2	1	6	my family
1	1	6	my family (especially my wife)
2	1	2	happiness of family as a whole
2	3	2	health and good living
1	2	4	family life, unity of family life, we have a good life here, very comfortable, very easy, understanding children, not ignoring them, joining in etc.
2	1	6	well I suppose just living, living and eating

First the words are sorted by their frequency in the entire corpus. Then they are sorted in alphabetical order within each frequency. We show an example of the coding of a corpus of responses to an open question (***Life*** corpus).

Then we introduce a corpus ***P*** (as in pedagogical) of drastically reduced dimensions, by way of introducing several concepts and definitions.

2.2.1 Numeric coding of *Life* corpus

In the corpus comprising *1000* responses to the ***Life*** open question, an excerpt of which is shown in table 2.1, the word *my* is the most frequent with *809* occurrences.

Table 2.2
Numeric coding of 5 responses to *Life* question presented in table 2.1

Gender	*Educ.*	*Age*	*Numeric codes*
1	1	4	11 12 40 187 69, 238 2, 113 137 69 18
1	2	2	1 138 72, 49 848 265 66 40
1	2	2	60 7 503 31 4 99 10 50 12 1 125 72
1	3	2	10 95 21
1	2	1	528, 1 21, 131 43 1 417 ' 19 528

Table 2.2 shows the numeric coding of the first responses presented in table 2.1. Thus numbering textual sequences has the result of simplifying lexicometric processing. Additionally, the volume needed for storage in the computer is considerably reduced.

2.2.2 Corpus *P*

In order to provide simple illustrations of the definitions given in this chapter, we work on a text called "corpus *P*", where capital letters are used to represent words whose exact style and meaning are of no significance. However this simple "text" does provide some examples of redundancies that we shall use later on. This corpus has thirteen *occurrences* numbered *1* to *13*; its size is designated by the letter *T*.

Table 2.3
Pedagogical corpus *P*

A	B	C	A	C	;	C	D	E	A	B	C .	D	F.
1	2	3	4	5		6	7	8	9	10	11	12	13

Among these occurrences, some are occurrences of the same word. Thus, occurrences *1*, *4* and *9* are all occurrences of the word *A*. Six different words are counted in total in this corpus: *A, B, C, D, E, F*. These words constitute the vocabulary of corpus *P*. Designating by *V* the vocabulary count, we write here: $V = 6$.

2.3 QUANTITATIVE ANALYSIS OF VOCABULARY

2.3.1 Frequencies, frequency distribution

Returning to the corpus *P* example, word *A* has three occurrences: its frequency is equal to three, and the *addresses* of this word in the corpus are:

1, 4 and *9*. In this corpus, word *C* is the most frequent word, with *4* occurrences; the *maximum frequency* is thus equal to *4*. This is usually noted:

$$F_{max} = 4$$

Word *E* only appears once in the text; it is a word of frequency *1*, also known as a *hapax*[1]. In this corpus there are two hapaxes : words *E* and *F*. Designating by V_1 the count of words of frequency *1*, we write:

$$V_1 = 2$$

Counts corresponding respectively to frequencies two, three and four are written:

$$V_2 = 2;\ V_3 = 1;\ V_4 = 1.$$

Frequency series can also be represented as follows:

Frequency:	1	2	3	...	F_{max}
Count:	V_1	V_2	V_3	...	$V_{F_{max}}$

For corpus *P* the frequency series is written:

Frequency:	1	2	3	4
Count:	2	2	1	1

The frequencies and their associated counts are linked with one another through overall relationships, some of which are very simple. The sum of the counts across all frequencies is equal to the number of words contained in the corpus, which is written:

$$\sum_{i=1}^{Fmax} V_i = V$$

The sum of the (frequency x count) products for all frequencies between *1* and F_{max} boundaries included, is equal to the length of the corpus.

$$\sum_{i=1}^{Fmax} V_i \times i = T$$

2.3.2 Zipf's law

If the corpus being processed has several thousand occurrences, it has been observed that the frequency distribution acquires characteristics that are more complex. These characteristics have been associated with the name of G.K. Zipf. Zipf's first discovery[2] was the observation that all of the frequency distributions obtained from textual corpora have characteristics in common. Roughly speaking, if the elements of a frequency distribution are sorted in

1 From the Greek *hapax legomenon*, "something stated once".

2 Mandelbrot (1961), emphasizes contributions of Estoup (1916) in this field, in a communication that discusses several attempts to model frequency series.

decreasing order, and then numbered, there is an approximate relationship between the rank and the frequency. Sometimes the following statement is made:

The product of rank and frequency is approximately constant.

Thus in Joyce's *Ulysses*, for example, which has *260 000* occurrences, counts performed in 1937 by H.L. Handley, yield:

rank	*10*	*frequency*	*2653*
rank	*100*	*frequency*	*265*
rank	*1000*	*frequency*	*26*
rank	*10 000*	*frequency*	*2*

A Pareto chart furnishes a very good summarization of the information contained in frequency distribution. On the vertical axis, with rulings in a logarithmic scale, the frequency of repetition F, which varies from *1* to F_{max} maximum frequency of the corpus, is entered. On the horizontal axis, again in a logarithmic scale, the number $N(F)$ of words repeated at least F times in the corpus is entered for each of the values of frequency F between *1* and F_{max}. The resulting curve is a cumulative curve.

Many experiments performed in the field of lexicometrics have shown that regardless of the texts considered, and regardless of the processing norm utilized, the points thus drawn align themselves approximately in a straight line.

Figure 2.1 shows an example of a Pareto chart comparing text S_1 with corpora formed by juxtaposing responses to open-end questions. S_1 is a fragment of *20 000* occurrences drawn from labor-union texts. Q_1 is a text comprising the concatenation of *2000* responses to an open ended question in a socio-economic survey (corresponding to *15 000* occurrences).

The chart can be used to characterize the frequency distribution of text Q_1 compared to text S_1. On figure 2.1, it can be seen that the maximum frequencies (i.e. frequency of the most frequent word) are close for the two texts even though Q_1 has *5 000* occurrences less than S_1.

Guide to reading Pareto chart

The frequency of repetition F is read on the vertical axis, with rulings in a logarithmic scale.
On the horizontal axis, scaled in the same way, is found the number of words repeated at least F times in the corpus, for each frequency F between *1* and F_{max}.

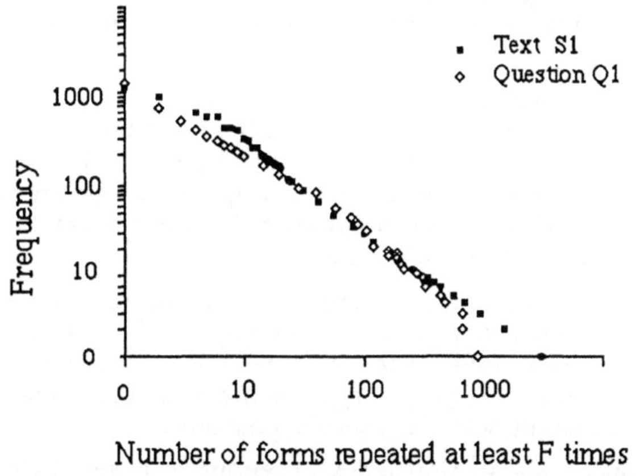

Figure 2.1
Pareto chart for "texts" *S_1* (*20 000* occ. of continuous text) and *Q_1* (*15 000* occ. from responses to an open question)

Words with small frequencies are more numerous in text S_1. This difference lessens as we approach a frequency of *20*. Between a frequency of *20* and a frequency of *100*, the curve for text Q_1 is above the curve for text S_1. This translates into a larger number of forms of moderate frequency for question Q_1. Between a frequency of *100* and the maximum frequency, the curve for text S_1 takes the lead, meaning that there is a relatively larger number of words whose frequency is within this interval compared to question Q_1. We conclude that text Q_1 has a more restricted vocabulary and greater repetitiveness in the range of moderate frequencies than does text S_1.

2.4 LEXICOMETRIC DOCUMENTS

Indexes and concordances are simply rearrangements of the textual sequence in a corpus of texts. Their value lies in their ability to provide a perspective that is different from that gained from a sequential reading of the corpus[1].

---

1 Note that the practice of producing such documents is much older than computer-assisted lexicometric analysis.

2.4.1 Index of a corpus

An index is a rearrangement of the words and occurrences within a text. It is used to rapidly find every location in the corpus where occurrences of any word are situated.

To locate these occurrences in the initial text, a system of numeric coordinates that are closely related to the text's edition is used: volume, page, line, position of occurrence in the line, etc. This information, intended to facilitate referring to the original document, consists of *references* associated with each occurrence. Within indexes, words can be arranged according to different criteria.

An *alphabetic index* is an index where words are arranged in lexicographic order (i.e. the alphabetical order commonly used in dictionaries). An *hierarchical index* is an index where the first sorting criterion is decreasing frequency of occurrence. In this type of index, words having the same frequency are generally sorted in lexicographic order.

Table 2.4 shows a hierarchical index of corpus *P*. In this index the order number in the corpus serves as the reference for each of the occurrences.

Table 2.4
Hierarchical index of corpus *P*

frequency	*word*	*references*			
4	C	3	5	6	11
3	A	1	4	9	
2	B	2	10		
2	D	7	12		
1	E	8			
1	F	13			

2.4.2 Contexts, concordances

For each of the words of the corpus, the set of its occurrences in the original text can be located by using indexes. Once this work is accomplished, the immediate contexts from which these occurrences are extracted can be systematically examined. However, if a word has a high frequency, this operation, requiring a constant shuttling between the text and the index, rapidly becomes computationally costly.

However both words and their contexts within a text can be rearranged: occurrences of a given word can be congregated along with a small fragment of their immediate context, whose length is fixed according to analytical requirements. This process produces a *concordance*.

A *pivotal word* is a word whose contexts are shown together. Table 2.5 is an example of a concordance that brings together all occurrences relative to a single pivotal word, word C of corpus *P*.

Table 2.5
Concordance of form *C* in corpus *P*

preceding context	*pivotal-word*	*following context*	*reference*
A B	**C**	A C ;C D E A B	3
A B C A	**C**	; C D E A B C .D	5
A B C A C;	**C**	D E A B C .D F	6
C D E A B	**C**	. D F.	11

(reminder of corpus P) :

A B **C** A **C** ; **C** D E A B **C** . D F.

In the example of table 2.5, the rows of the concordance are sorted in alphabetical order of the word following the pivot (*A, C, D, D* in succession). As above, the last column is the occurrence number in the text.

Table 2.6
Concordance of *wife* in corpus of responses to *Life* question

the children well and decently	wife	and I live happily together
nt, don' t really mind that my	wife	and I remain as active as possible
I think that' s basically it,	wife	and children, thing my daughters a
wn life so I can look after my	wife	and daughter, being able to look a
w collecting records,(music),	wife	and daughter, to see a true social
amily happiness good health my	wife	and family good health health look
being of family and friends my	wife	and family
permanent job, my	wife	and family my grandchildren
ldren wealth and happiness, my	wife	and family my work, social life,
	wife	and family paying my way
my youngest son my	wife	and family to be happy change,
good health,	wife	and family work satisfaction, leis
my children my	wife	, and good health that I should kee
my	wife	and her happiness to be honest
rl money, health, happiness my	wife	and son, keeping my job
with every body, obviously the	wife	and the children, protected life
my grandson, his	wife	, and their son, he' s so precious
ry important to enjoy yourself	wife	, baby, health,
I would like to have my	wife	back, she died recently
health my	wife	family health my children' s well
my children, my	wife	family life and my home, sensible
just now welfare of children,	wife	, family, maintain, good standard
ion do as you would be done by	wife	, family, welfare of the community

In spite of their volume, concordances have revealed themselves to be very practical instruments for analyzing texts. Because they rearrange the textual sequence, concordances provide an overall perspective of the ways in which a word is used that would be difficult to achieve from sequential reading. In particular they provide an opportunity to more easily examine relationships that may exist among the various contexts of a word.

Table 2.6 shows some rows of a *concordance* performed on the contexts of the word *wife* in the set of responses to open question ***Life***. The context rows are sorted here in ascending alphabetical order on the word that follows the pivotal-word[1].

Note also that a sort on the occurrence preceding the pivotal word would have the effect of dispersing occurrences of these expressions to different locations in the concordance, and bringing together others that are currently dispersed by the sorting method used here.

2.4.3 Vocabulary growth

When occurrences are added to a text, its vocabulary (the number of distinct words) tends to grow. The study of the growth of the vocabulary as occurrences increase in a given text constitutes one of the traditional fields of lexical statistics.

For corpus *P*, this growth is described in the following manner:

	A	B	C	A	C	;	C	D	E	A	B	C	.	D	F.
occurrences	1	2	3	4	5		6	7	8	9	10	11		12	13
words	1	2	3	3	3		3	4	5	5	5	5		5	6

In a large corpus, vocabulary growth is subject to a double influence.

a) Taking new occurrences into account tends to increase the total number of words in the corpus (the more occurrences there are, the more numerous are distinct words).

b) As the size of the corpus increases, the rate at which new words are brought in by each increase in the number of occurrences tends to decrease.

It follows that the number of words contained in a corpus is not proportional to its size.

Among the variables that are used to characterize a text, we can make a distinction between type *T* lexicometric variables and type *V* lexicometric variables.

---

1 Note that this rearrangement of occurrences in the initial text uncovers associations that might have been missed in a simple sequential reading of the text

- ***type T*** lexicometric variables grow *approximately* in proportion to the length of a text.
- ***type V*** lexicometric variables, such as the number of words in the text, whose growth tends to become less as the length of the text increases.

2.4.4 Lexical tables

It is convenient to arrange counts of occurrences of each of the *V* words in each of the *N* parts of the corpus into a rectangular table with *V* rows and *N* columns. In this table, words are organized from top to bottom in lexicometric order (cf. table 2.7).

Table 2.7

Lexical Table (LT) of corpus *P* with partitioning into 3 parts and marginals (frequencies of words)

word	P1	P2	P3	*Total*
C	2	2	0	4
A	2	1	0	3
B	1	1	0	2
D	0	1	1	2
E	0	1	0	1
F	0	0	1	1
length of parts	5	6	2	13

This table is called: *Lexical Table* (LT). The numbers *V* and *N* are the dimensions of the LT. Partial lexical tables can be built by including only words whose frequency is at least equal to a certain threshold, for example.

The partitioning into three parts of corpus *P* can be used to build a LT of dimensions (6×3). The numbers constituting the vertical marginal of the table indicate the frequency of each word in the corpus [1].

2.5 REPEATED SEGMENTS

Even after setting aside words with a purely grammatical role, the meaning of words is linked to how they appear in compound words, or in phrases and expressions that can either inflect or completely change their meanings.

---

[1] In chapter 3 (table 3.1) and mostly in chapter 5 (table 5.3) are found examples of lexical tables built from survey data.

For example, expressions such as: *social security, living standard, etc.,* have a meaning of their own that cannot be construed from the meaning of the words of which they are composed. It is thus useful to count larger units consisting of several words. These elements can be analyzed in the same ways as words. But decisions concerning how to group occurrences of words cannot be made without bringing subjectivity into the analytical process. In this section we introduce a new type of textual object: *repeated segments.*[1] In subsequent chapters we shall demonstrate how these formal objects are used with different statistical methods[2].

2.5.1 Sentences, sequences

Before undertaking descriptions of segments consisting of several words and repeated within a corpus of texts, it is useful to specify a status for each delimiting character.

When series of words are counted without taking into account punctuation marks that might be interspersed within them, two series such as: [A,B,C D] and [A B C D] are counted as two occurrences of the same segment. This might become a source of confusion.

To avoid having to perform separate counts for occurrences of segments containing punctuation marks and segments that are identical except for punctuation, the usual practice is to search for segments that do not straddle weak or strong punctuation marks. Among linguists, the problem of defining the limits of a sentence has the reputation of being a difficult one. Unfortunately, a definitive solution has not been found for this problem within the lexicometric field. Here we choose to assign the status of strong separator or sentence separator to some of the punctuation marks (period, exclamation point, question mark).

We return to the example of corpus **P** presented above:

A	B	C	A	C	;	C	D	E	A	B	C	.	D	F.
1	2	3	4	5		6	7	8	9	10	11		12	13

We observe that corpus **P** is composed of two sentences. We also choose a subset of weak punctuation marks (usually: comma, semicolon, colon, hyphen, quotation marks and parentheses). All of these delimiting characters together are called sequence delimiters.

---

1 Cf. Lafon and Salem (1983), Salem (1984, 1987). Cf. also Becue (1988).

2 Numerous dictionaries of phrases, compound words, are available in the domain of corpus linguistics ; for example: in English (Fagan, 1987); in French (Silberztein, 1993; Gross and Perrin, 1989); in Italian (Elia, 1995). The repeated segments approach is significantly different : it deals with the detection of repeated sequences of words within a given corpus, whether or not they constitute frozen phrases or expressions.

Word delimiters that are neither sentence separators nor sequence delimiters are treated as characters called blanks or spaces. We see that corpus **P** comprises three sequences. The classification of characters derived above is summarized in figure 2.2.

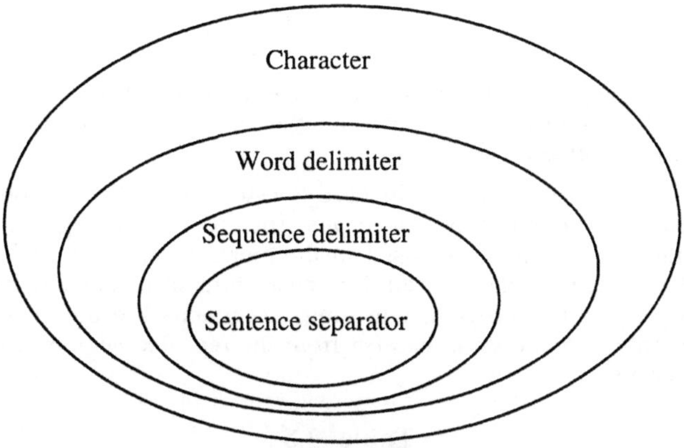

Figure 2.2
Classification of characters

According to the above definitions, all series of occurrences not separated by a sequence delimiter are occurrences of segments. Segments whose frequency is greater than or equal to 2 in the corpus are *repeated segments.*

We return to corpus **P**. The first ABCAC sequence of corpus **P** can be broken up into segments in the following manner:

ABCAC, ABCA, BCAC, ABC, BCA, CAC, AB, BC, CA, AC.

For corpus **P** as a whole, only segments ABC, AB, BC are repeated segments. The segment of length 3, ABC, is a repeated segment of the corpus. It is also true that any sequence of length *L* necessarily contains: *(L-1)* segments of length 2, *(L-2)* segments of length 3, etc.

Constrained segments, free segments

Sub-segments of a given segment are divided in the same way into two categories: *constrained segments* and *free segments*. A sub-segment is constrained (in another segment *S* of greater length) if all of its occurrences happen within occurrences of segment *S*.

2.5.2 Repeated segments table

Repeated segment tables are tables in which are displayed counts relative to the distribution of occurrences of a segment in different parts of a corpus. Later we shall see how these tables are used in various statistical analyses of repeated segments.

For corpus **P** the repeated segments table is thus written as follows:

	part		
segment	P1	P2	P3
ABC	*1*	*1*	*0*

All the other segments of corpus **P** are eliminated from this repeated segments table either because they are hapax segments, or because they are constrained in segment *ABC*.

The repeated segments table can be used to estimate the distribution of each segment in the different parts of a corpus. The order of the segments in this table is the same as the order used in the hierarchical index of forms (cf. section 2.4.1). This table is used for most statistical analyses based on segments. Segments constrained in longer segments are eliminated from this table. Table 2.8 shows an excerpt from the repeated segments table for the ***Life*** corpus.

Table 2.8

Distribution of some frequent segments within 3 parts (*Life* open question)

segment	P1	P2	P3
family and friends	*5*	*4*	*6*
health and happiness	*8*	*8*	*1*
to be able to get	2	*1*	*1*
to do what I want	*1*	2	*1*
to have good health	*3*	*1*	*0*
to live comfortable	*5*	*5*	*1*
good family life	*4*	*1*	*0*
good standard of life	*5*	*4*	*1*
money to live on	*4*	*6*	*0*
peace in the world	2	*3*	*1*
peace of mind	*11*	22	9
having enough money	*4*	*8*	*1*
enough money to keep	*0*	*4*	*1*
no money worries	*3*	*1*	*0*
keeping my job	2	2	*0*
standard of living	*10*	*14*	6
making sure	*3*	*4*	*0*
old age	2	2	*0*
helping people	2	2	*0*
quality of life	2	*4*	2

2.6 FINDING CO-OCCURRENCES, QUASI-SEGMENTS

Research on repeated segments grew out of difficulties encountered in the search for co-occurrences (special attractions between pairs of words within a given context). Without claiming to be exhaustive, since the methods used to this end vary considerably as a function of the fields of application, we now discuss some of these methods.

2.6.1 Searching around a pivotal word

Given a pivotal-word, several methods can be used to select the set of words that tend to appear often in the neighborhood of this word. In order to select these words, a unit of context or neighborhood must be chosen, within which two words are considered to be co-occurring. This unit can be similar to a sentence or can consist of a fixed-length context (x occurrences before and x occurrences after the pivotal-word).

For each unit *word1* of the corpus, a simple method of defining the *lexical universe* of a given word consists in dividing the set of sentences of the corpus into two subsets: P_1, subset of those that contain *word1* and P_0, subset of sentences from which *word1* is absent. For each of the other words of the corpus, statistical tests are performed on the sub-frequencies in the two sets P_0 and P_1 taking into account their respective lengths. If the frequencies of the words under consideration are not too small, the method can be used to select, for each pivotal word, a set of words that are positioned preferentially in the same sentences[1].

Generally, the search for preferential associations is an important factor in applications involving *natural language processing*. Thus some ambiguities encountered in optical scanning of characters can be removed (at least in terms of probability) by considering neighboring words that have already been recognized, if the probabilities of association are known.

Disambiguation in the course of a morpho-syntactic analysis can be achieved under the same conditions. The work of Church and Hanks (1990) belongs to the same general framework. These authors suggest using as a measure of association between two words x and y the mutual information index $I(x,y)$ resulting from the Information theory of C. Shannon:

$$I(x,y) = \log_2 \frac{P(x,y)}{P(x)P(y)}$$

1 Another *method of co-occurrences*, (Lafon, 1981) assigns a probability index to the number of encounters of each pair of words within a sentence (sequence of words between two strong punctuation marks). This method makes it possible to select, using a probability threshold, sets of pairs of words that show affinities in the text.

where $P(x)$ and $P(y)$ are the frequencies of words x and y in a corpus, and $P(x,y)$ is the frequency of neighboring occurrences of these two words; x precedes y (thus there is no symmetry with respect to x and y), and closeness is defined by a distance counted as a number of words. Thus, for English language texts, these authors recommend considering as neighbors two words separated by fewer than five words.[1]

2.6.2 Finding multiple co-occurrences, quasi-segments

The methods whose objective is to find co-occurrences of two or several words in a text have been developed particularly in information retrieval. Procedures that select repeated segments in a corpus are not yet able to find repetitions that are slightly altered by the introduction of an adjective or by a minor lexical modification of one of the components. Becue (1993) proposed an algorithm that is used to find (repeated) *quasi-segments*. This algorithm can be used, for example, to assemble into one element *(play, sport)* the sequences *play a sport* and *play some sports*. However quasi-segments are even more numerous than segments, and counting them gives rise to problems in their selection and display (see also chapter 4, section 4.2.3).

2.7 DEALING WITH TAGGED CORPORA

2.7.1 Appending metadata to free responses in surveys

Automatic tagging of the words of a corpus of responses provides an indication of the lemma to which each word can be associated as well as its grammatical category. In most cases the information furnished by categorizers can be subjected to careful checking before being used, since the process of automatic categorization and lemmatization generates many errors.

Table 2.9 shows an example of a grammatical categorization for the first ten responses of the ***Life*** corpus[2]. An example of simultaneous processing of words and syntactic annotations (in the case of responses to an open-ended question) has been performed by Habert and Salem (1995).

In what follows we choose most often to illustrate the quantitative methods we present on the basis of counts performed on words. As we shall see,

---

1 This procedure seems efficient for identifying some *phrasal verbs*. In a corpus of *44* million occurrences, the authors show that the verb *set* is followed first by *up* $(I(x,y)=7.3)$ and then by *off* $(I(x,y)=6.2)$, then by *out, on, in,* etc.

2 The main grammatical codes used in this categorization are:
NN - Noun singular; NNS - Noun plural; NP - Proper noun;
IN - Preposition or subordinating conjunction; RB - Adverb; DT - Determiner;
VB - Verb, base form; VBD - Verb, past tense; VBG - Verb, gerund or present part;
VBN - Verb, past participle; MD - Modal;
JJ - Adjective; PRP - Personal pronoun; PRP$ - Possessive pronoun; etc.

these methods can be applied without modification to counts performed on lemmas or on categorized elements[1].

Table 2.9
Examples of responses to *Life* question with grammatical category code for each word

Gender	*Educ.*	*Age*	***Tagged responses***
1	1	4	happiness/NN in/IN people/NNS around/IN me/PRP ,/, contented/VBN family/NN ,/, would/MD make/VB me/PRP happy/JJ
1	2	2	my/PRP$ own/JJ time/NN ,/, not/RB dictated/VBN by/IN other/JJ people/NNS
1	2	2	freedom/NN of/IN choice/NN as/IN to/TO what/WP I/PRP do/VB in/IN my/PRP$ leisure/NN time/NN
1	3	2	I/PRP suppose/VBP work/NN
1	2	1	firm/NN ,/, my/PRP$ work/NN ,/, which/WDT is/VBZ my/PRP$ dad's/NNS firm/NN
2	1	6	just/RB the/DT memory/NN of/IN my/PRP$ last/JJ husband/NN
2	2	6	wellbeing/NN of/IN my/PRP$ handicapped/JJ son/NN
1	1	5	my/PRP$ wife/NN ,/, she/PRP gave/VBD me/PRP courage/NN to/TO carry/VB on/IN even/RB in/IN the/DT bad/JJ times/NNS
2	2	3	my/PRP$ sons/NNS ,/, my/PRP$ kids/NNS are/VBP very/RB important/JJ to/TO me/PRP ,/, being/VBG on/IN my/PRP$ own/JJ I/PRP am/VBP responsible/JJ for/IN their/PRP$ education/NN
1	3	3	job/NN ,/, being/VBG a/DT teacher/NN I/PRP love/VBP my/PRP$ job/NN ,/, for/IN the/DT wellbeing/NN of/IN the/DT children/NNS

2.7.2 Comparison of main quantitative characteristics

To conclude this chapter we present a comparison of counts performed on both graphical forms (words) and lemmas, on a relatively lengthy corpus.

This comparison is useful to measure the effect of lemmatization on lexicometric counts. Later we shall analyze (Chapter 7, section 7.2) the

[1] Note that the way of tagging a corpus may differ according to the purpose of the investigation. Cf. Biber (1988), Brill (1992), Cutting (1992), Oostdijk (1991).

consequences of changing the unit of counting on statistical descriptions concerning the parts of this corpus (***Speeches*** corpus).

The ***Speeches*** corpus is a collection of 68 presidential speeches of F. Mitterrand that took place between July 1981 and March 1988.[1]

Table 2.10 can be used for a first comparison between counts performed on lemmas and those done on the graphical forms of the corpus.

Table 2.10

***Speeches* corpus : main lexicometric characteristics of processing as graphical forms and as lemmas**

	forms	*lemmas*
number of occurrences	*297 258*	*307 865*
number of words	*13 590*	*9 309*
number of hapaxes	*5 543*	*3 255*
maximum frequency	*11 544*	*29 559*

As can be seen on this table, the number of occurrences is higher for the lemmatized corpus (+*10 607* occurrences). This circumstance is not unduly surprising because there are two broad types of operations that have opposite effects on the number of occurrences in a corpus for a lemmatization of the type considered above:

- lemmatization of certain units composed of several words (*at least*) tends to reduce the number of occurrences of the lemmatized corpus.
- conversely, separating each of the numerous occurrences of contracted graphical forms (*cannot*) into several distinct elements tends to increase the number of occurrences of the lemmatized file as compared to the initial text.

The large excess in the number of *occurrences* observed for the lemmatized file compared to the original text tends to prove simply, as we might have expected, that the second phenomenon described above is clearly more influential than the first.

Analogous reasoning can be used to explain the fact that the total number of *words* of the lemmatized corpus is clearly smaller than in the non-lemmatized corpus. Here again two phenomena come into play in opposite directions in the lemmatization process:

---

[1] The series of textual data we use in this section was assembled by D. Labbé (1990). This set includes texts of speeches made during the first seven year term and a lemmatization file indicating for each occurrence in the text its assigned lemma and its categorization.

- grouping several graphical forms that correspond to different inflections of the same lemma (verb inflections, singular and plural of nouns, etc.) reduces the number of words in the lemmatized corpus.
- disambiguation through identifying the grammatical category of each of the words (*book* -verb different of *book* -noun) tends to increase this number.

From a quantitative point of view, it can be seen that the process of grouping the graphical forms prevails in a lemmatization. This also explains why the number of hapaxes in the corpus is also far smaller than its equivalent measured in terms of words.

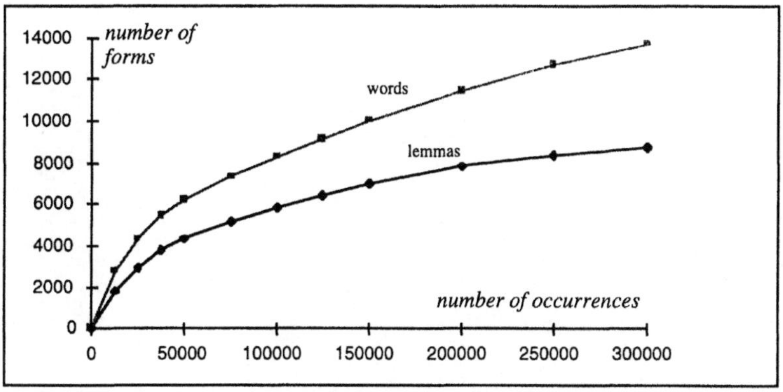

Figure 2.3

Vocabulary growth measured as graphical forms (words) and as lemmas

Measure of vocabulary growth

Figure 2.3 shows vocabulary growth measured on the two types of units. As can be seen, the general shapes of the growth curves are similar although the curve for the lemmatized corpus is lower. These findings imply that comparisons relative to major lexicometric characteristics should be limited to counts performed using identical processing norms.

3

CORRESPONDENCE ANALYSIS OF LEXICAL TABLES

Cross-tabulations or contingency tables are among the most common data structures used for analyzing qualitative data. By looking simultaneously at two partitions at a time of a population or sample, a cross-tabulation enables us to work with *variations in the data by response categories*, a necessary step for the initial interpretation of results. Furthermore, analyses and descriptions of cross-tabulations are the basis for the statistical treatment of survey data.

The main purpose of multidimensional descriptive data analysis methods is to describe tables such as tables of measures, classification tables, incidence tables, contingency tables or presence-absence tables. These methods have profoundly changed the way survey data is analyzed by providing analysts with global panoramas by themes, tools that test the coherence of the information and procedures for selecting the most relevant cross-tabulations.

Visualization tools (principal axis planes) and clustering algorithms (methods that group individuals by simultaneously taking into account several responses or several basic characteristics) make it possible for the analyst to gain a macroscopic view of the raw data.

The lexical tables to which we plan to apply these techniques are special contingency tables: the statistical individual being counted in each cell of the table is the *occurrence* (or *token*) of a textual unit: word, lemma, repeated segment, etc. The rows of the table are for example words whose frequency in the data is greater than a given threshold, and the columns are groupings of text: speakers, categories of speakers, authors, documents, etc. In general cell (i,j) contains the number of occurrences of word i in text group j.

This chapter briefly reviews the basic principles and practicalities of using these methods by way of an introduction to discussing their special contribution to new technologies for survey processing.

3.1 BASIC PRINCIPLES OF MULTIVARIATE DESCRIPTIVE METHODS

The common principle to all exploratory multivariate statistical methods is the following:

Each one of the two dimensions of a rectangular table of numerical data can be used to define distances (or proximities) among the elements of the other dimension. Thus the table's columns (which might be variables, attributes, etc.) are used to define the distances among the table's rows which might be individuals, observations, etc.) with the help of appropriate formulas. Similarly the table's rows are used to calculate distances among the table's columns.

In this fashion are obtained tables of distances. These tables are the source of complex geometric representations that describe the similarities among the rows and among the columns of the rectangular tables being analyzed. The problem is to make these visualizations as understandable and as intuitive as possible, while keeping at a minimum the loss of information that results from summarizing the data.

Briefly, there are two broad types of methods for performing these data reductions:

- *Principal axes methods,* largely based upon linear algebra, produce graphical representations on which the geometric proximities among row-points and among column-points translate statistical associations among rows and among columns. Correspondence analysis, presented in this chapter, belongs to this family of methods.
- *Clustering or classification methods* that create groupings of rows or of columns into clusters (or into families of hierarchical clusters) are presented in the following chapter.

These two families of methods can be used on the same data matrix and they complement one another very effectively. However it is obvious that the rules of interpretation for the representations obtained through these reduction techniques are not as simple as those of elementary descriptive statistics. A rudimentary level of training is sufficient for interpreting histograms, bar charts and time series graphs. In the case of correspondence analysis, for example, it is necessary to be familiar with rules of interpretation of the results. These rules are not straightforward, despite the provocative appearance of the graphical displays. In addition to formal training, the analyst needs to have hands-on experience (by that we mean varied and informal experience, since each data matrix has its own peculiarities).

A pedagogical paradox

In order to understand how this method works, it is necessary to work on a small example, one that is easily grasped by the user. But the power of the techniques of multidimensional exploratory analysis comes from their ability to analyze very large data sets. Perhaps the following image summarizes this pedagogical paradox: it is extremely difficult to demonstrate the effectiveness of a fish net in a living room aquarium!

We shall nevertheless show how these methods work on small data matrices to illustrate the principles without initially demonstrating these tools' full effectiveness. The examples in chapter 5 will undoubtedly do a better job of showing their real power in data exploration and summarization.

3.2 CORRESPONDENCE ANALYSIS

Correspondence analysis is a technique for describing contingency tables (or cross-tabulations) and certain binary tables (also known as "presence-absence" tables). This description essentially takes the form of a graphic representation of associations among rows and among columns.

3.2.1 Brief historical overview

Correspondence analysis was first presented in a systematic way as a flexible technique for exploratory analysis of multidimensional data by Benzécri (1973). A disparity of sources, all independent from one another are either precursors to correspondence analysis or involved in further research. Among them we mention Guttman (1941) and Hayashi (1956), both of whom mention *quantification methods*, Nishisato (1980) who speaks of *dual scaling*, and Gifi (1990), who speaks of *homogeneity analysis* to designate multiple correspondence analysis.[1]

3.2.2 Correspondence analysis - a simple numerical example

Table 3.1 is a frequency table. In this table the *14* rows are words used in responses to an open-ended question given by *2000* respondents[2]. The five columns are the educational levels[3] of the respondents. Recall that the statistical unit here is not the respondent but the occurrence of a word. The columns constitute a partitioning of the set of respondents but the rows do not: a single respondent can use several words in the list within his or her

[1] Cf. overviews by Tenenhaus & Young (1985) and Escoufier (1985), English presentations and examples are found in Greenacre (1984, 1993), Lebart *et al.* (1984) and Benzécri (1992c), Gower and Hand (1996).

[2] "*What are the reasons that might cause a couple or a woman to hesitate having children ?*" Survey about aspirations and life styles of the French.

[3] No degree, Elementary School degree, Trade School degree, High School degree, College degree.

response. The rows constitute a partitioning of the set of occurrences of the words.

Table 3.1
Cross-tabulation of words with Educational level. Raw frequencies

Words	No degree	Elem. Sch.	Trade Sch.	High Sch.	College	Total
Money	51	64	32	29	17	193
Future	53	90	78	75	22	318
Unemployment	71	111	50	40	11	283
Decision	1	7	5	5	4	22
Difficult	7	11	4	3	2	27
Economic	7	13	12	11	11	54
Selfishness	21	37	14	26	9	107
Occupation	12	35	19	6	7	79
Finances	10	7	7	3	1	28
War	4	7	7	6	2	26
Housing	8	22	7	10	5	52
Fear	25	45	38	38	13	159
Health	18	27	20	19	9	93
Work	35	61	29	14	12	151
Total	323	537	322	285	125	1592

The table is read as follows: the word *Money* for example, was used *51* times by persons belonging to the category "no degree". The row totals represent the number of occurrences of each word whereas the column totals represent the total number of words (within the list) used by the various categories of respondents.

Table 3.2
Row-profiles of table 3.1

Words	No degree	Elem. Sch.	Trade Sch.	High Sch.	College	Total
Money	26.4	33.2	16.6	15.0	8.8	100.0
Future	16.7	28.3	24.5	23.6	6.9	100.0
Unemployment	25.1	39.2	17.7	14.1	3.9	100.0
Decision	4.5	31.8	22.7	22.7	18.2	100.0
Difficult	25.9	40.7	14.8	11.1	7.4	100.0
Economic	13.0	24.1	22.2	20.4	20.4	100.0
Selfishness	19.6	34.6	13.1	24.3	8.4	100.0
Occupation	15.2	44.3	24.1	7.6	8.9	100.0
Finances	35.7	25.0	25.0	10.7	3.6	100.0
War	15.4	26.9	26.9	23.1	7.7	100.0
Housing	15.4	42.3	13.5	19.2	9.6	100.0
Fear	15.7	28.3	23.9	23.9	8.2	100.0
Health	19.4	29.0	21.5	20.4	9.7	100.0
Work	23.2	40.4	19.2	9.3	7.9	100.0
Total	20.3	33.7	20.2	17.9	7.9	100.0

Table 3.2 shows the row-profiles expressed as percentages; they are obtained by dividing each element of the table by the corresponding row-sum: the row-profile of the word *Fear*, for example, is obtained by dividing each number in that row by *159*.

By comparing two row-profiles, we learn how the words represented by these two profiles are associated with the categories (columns)[1].
This comparison is more difficult if we use table 3.1 alone, because the frequencies of the words vary a lot. Thus it is not immediately obvious in table 3.1 that *Decision* is used relatively often by college graduates. But this is easy to see in table 3.2.

Table 3.3 shows column-profiles: these are obtained in analogous fashion by dividing the elements of each column by their sum, and multiplying the result by *100*.

Table 3.3
Column-profiles of table 3.1

Words	No degree	Elem. Sch.	Trade Sch.	High Sch.	Coll-ege	Total
Money	15.8	11.9	9.9	10.2	13.6	12.1
Future	16.4	16.8	24.2	26.3	17.6	20.0
Unemployment	22.0	20.7	15.5	14.0	8.8	17.8
Decision	.3	1.3	1.6	1.8	3.2	1.4
Difficult	2.2	2.0	1.2	1.1	1.6	1.7
Economic	2.2	2.4	3.7	3.9	8.8	3.4
Selfishness	6.5	6.9	4.3	9.1	7.2	6.7
Occupation	3.7	6.5	5.9	2.1	5.6	5.0
Finances	3.1	1.3	2.2	1.1	.8	1.8
War	1.2	1.3	2.2	2.1	1.6	1.6
Housing	2.5	4.1	2.2	3.5	4.0	3.3
Fear	7.7	8.4	11.8	13.3	10.4	10.0
Health	5.6	5.0	6.2	6.7	7.2	5.8
Work	10.8	11.4	9.0	4.9	9.6	9.5
Total	100.0	100.0	100.0	100.0	100.0	100.0

By comparing two column-profiles we learn about similarities that exist among the various educational categories with respect to vocabulary used.

As illustrated in figure 3.1, the rows of table 3.2 can be viewed as 14 points in a 5-dimensional space, and the columns of table 3.3 can be viewed as 5 points in a 14-dimensional space.

When two words (rows of table 3.2) are represented by neighboring points, this means that the corresponding row-profiles are similar.

[1] To improve the readability of this table, the numbers have been multiplied by *100*.

Similarly, the closeness of two points representing categories of educational levels (columns of table 3.3) represents a similarity in column-profiles.

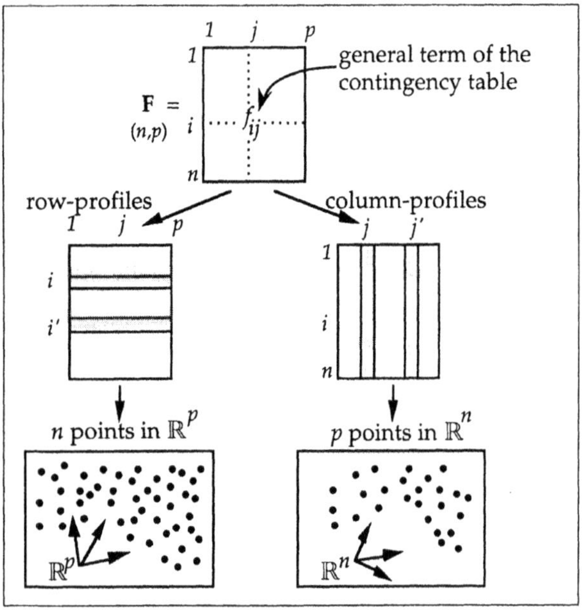

Figure 3.1
The two sets of points in two higher-dimensional spaces

Usual notation for profiles

Let us denote by f_{ij} the general term of the frequency table the elements of which have been divided by the grand total k (k=*1592* in table 3.1).

This table has n rows and p columns (in this case, n=*14* and p=*5*).

According to the usual notation, $f_{i.}$ designates the sum of the elements of row i and $f_{.j}$ is the sum of the elements of column j of this table.

The profile of row i is the set of p values:

$$\left(\frac{f_{ij}}{f_{i.}}\right), \quad j = 1,...p$$

The profile of column j is the set of n values:

$$\left(\frac{f_{ij}}{f_{.j}}\right), \quad i = 1,...n$$

The elements of the profiles multiplied by 100 are the row-percentages and column-percentages shown in tables 3.2 and 3.3.

Reduction of dimensionality

Since higher-dimensional spaces are difficult to inspect, it becomes necessary to reduce the dimensionality of the points. This reduction of dimensionality cannot be obtained without a certain loss of information. Principal axes methods (principal components analysis, factor analysis, correspondence analysis) aim to find the lower-dimensional subspace which most accurately approximates the original distributions of points.

Correspondence analysis and principal components analysis are used under different circumstances: principal components analysis (that can be viewed as a purely descriptive variant of factor analysis) is used for tables consisting of continuous measurements. Correspondence analysis is best adapted to contingency tables (cross-tabulations). By extension, correspondence analysis also provides a satisfactory description of tables with binary coding (see section 3.3).

Most of these methods provide the user with a sequence of nested subspaces. That means that the best one dimensional subspace (a straight line) is included in the best two-dimensional subspace (a plane) which in turn is included in the best three-dimensional subspace, etc.. In such a series of nested subspaces, the two-dimensional one is a very special case, since it is compatible with our most usual communication devices such as sheets of paper or video screens[1].

As a consequence, the following sections mainly deal with two-dimensional displays, which rarely give an exact representation of the distances between profiles, but are easier to inspect[2].

An example of a two-dimensional map

Figure 3.2 is a two-dimensional display generated by a correspondence analysis of table 3.1. It is a visual representation, or map, which simultaneously describes the similarities among row-profiles and similarities among column-profiles.

For someone who has a good grasp of the rules for interpreting such a map, this is a quick process for assimilating information. However one should not expect any major surprises or revelations from a graphical representation of a table of such modest dimensions. In this case correspondence analysis has a purely descriptive role: it simply makes results easier to look at.

---

[1] The term *biplot*, coined by Gabriel (1971), is sometimes used to designate various methods leading to two-dimensional displays of multivariate data, as exemplified by Gower and Hand (1996).

[2] It is easy to obtain good three-dimensional (interactive) displays on computer screen. It is always possible to complement the displays iteratively with the subsequent axes or planes.

Let us show, with the help of this example, how simple the principles and rules for interpretation of the method are.

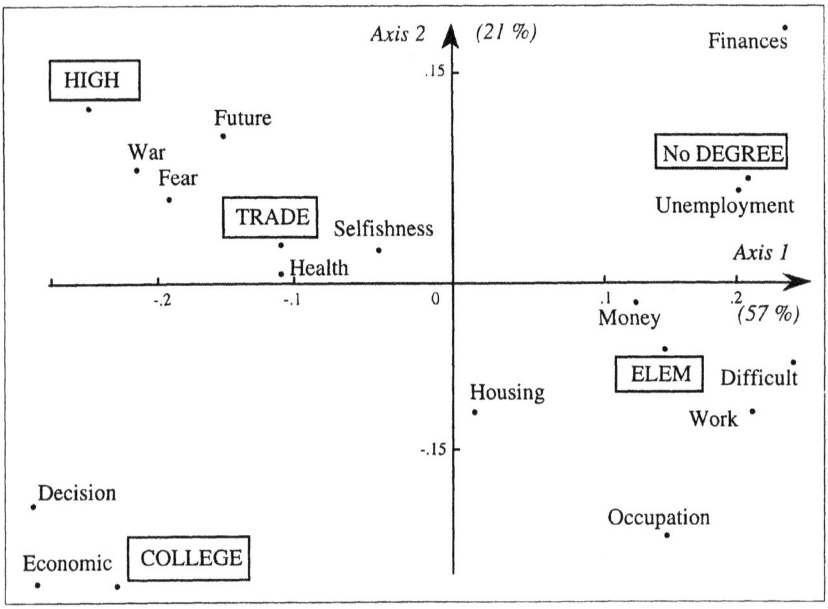

Figure 3.2
Proximities among words and among educational levels
Correspondence analysis of table 3.1
(the boxed elements correspond to the columns of table 3.1)

How does one read figure 3.2 ?

If two row-points i and i' have identical or similar profiles, they appear in exactly, or almost the same, position on each of the principal axes. In analogous fashion if two column-points j and j' have identical or similar profiles, they are in the same position or very close.

The origin of the axes represents the mean profiles (i.e., the marginals of the table of frequencies).

The components of the mean row-profile are: $f_{.j}$, $j=1,...p$

The components of the mean column-profile are: $f_{i.}$, $i=1,...n$

Thus, a column-point (boxed elements) such as ***Trade School***, which is rather close to the origin, has a similar profile to that of the *Total* column (the vertical marginal) of table 3.3. Similarly, a row-point such as *Health* has a profile that is similar to the *Total* row (horizontal marginal) which is the last row of table 3.2.

What is the meaning of the proximities?

We now define what is meant by "close". In order to do so, we explain how the distances whose planar approximations are shown in graphs such as figure 3.2 are calculated.

The profiles, which are series of n or p numbers, depending on whether we are dealing with rows or columns, are used to define points in spaces of p or n dimensions (more exactly, p-1 and n-1 dimensions, because the sum of the components of each profile is equal to 1).

Thus, the distances between points are defined in these high dimensional spaces. The analysis phase consists in reducing these dimensions to obtain a visual representation that distorts the distances as little as possible[1].

The distance between two row-points i and i' is given by the equation:

$$d^2(i,i') = \sum_{j=1}^{p} \frac{1}{f_{.j}} \left(\frac{f_{ij}}{f_{i.}} - \frac{f_{i'j}}{f_{i'.}} \right)^2$$

similarly, the distance between two column-points j and j' is given by:

$$d^2(j,j') = \sum_{i=1}^{n} \frac{1}{f_{i.}} \left(\frac{f_{ij}}{f_{.j}} - \frac{f_{ij'}}{f_{.j'}} \right)^2$$

This distance, which is called *chi-square distance,* is very much like the Euclidean distance (the sum of squares of differences between components of the profiles) but a weighting factor is also involved. This weight is the reciprocal of the frequency corresponding to each term: $(1/f_{.j})$ for each term in the sum that defines $d^2(i, i')$, and $(1/f_{i.})$ for each term in the sum that defines $d^2(j, j')$.

The chi-square distance has a remarkable property which is called *distributional equivalence*. Briefly, this property means that the distances among rows (or columns) remains the same when we merge two columns (or rows) that have identical profiles. This invariance property induces a certain stability of the results: thus two categories of educational levels that have the same lexical profile can be considered either separately or as one, without affecting the positions of the other points on the graph[2] .

Why a simultaneous representation?

We now know how to interpret the proximity between two row-points or two column-points, as well as their respective positions relative to the origin of the axes.

---

1 More technical details are given in Appendix 1.

2 For a more detailed explanation of the weighting procedure, cf. appendix 1. Cf. also: Benzécri (1973). Escofier (1978) presents several other distances having this property.

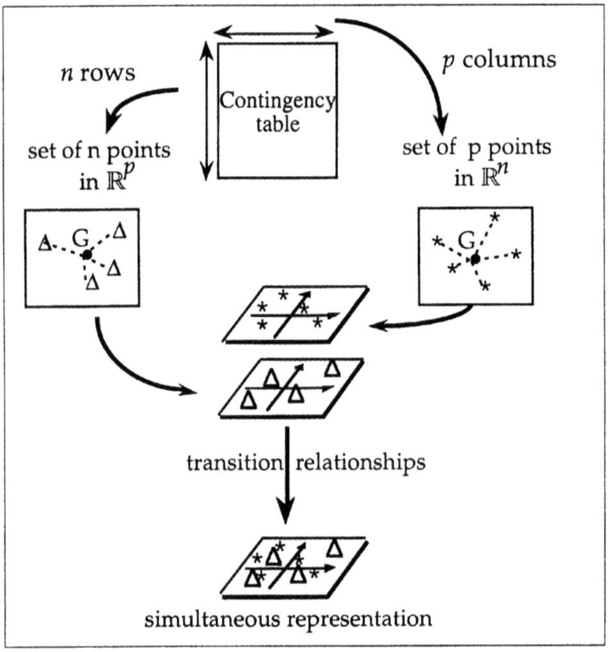

Figure 3.3
Simultaneous representation in correspondence analysis

But figure 3.2 shows us row-points and column-points simultaneously, and thereby displays additional proximities that we are tempted to interpret: it is not surprising that the row-point *Unemployment* should be close to the column-point ***No Degree***. But the proximity between *Health* and ***Trade School*** is less obvious.

As a matter of fact, it is not possible to interpret these cross-proximities between a row-point and a column-point, because these two points do not come from the same initial space. Nevertheless, it is possible to interpret the position of a single row-point with respect to the set of column-points or of a single column-point with respect to the set of row-points.

The main reason for this simultaneous representation is given by the *transition relationships* that link the coordinates of one point in one space (the row-space for example) to those of all the points of the other space (the column-space in our example).

If ψ_i designates the coordinate of a row-point i on the horizontal axis of figure 3.2 (first principal axis), and if φ_j designates the coordinate of a column-point j on the same axis, we have a remarkably symmetrical system of relationships:

$$\psi_i = \beta \sum_{j=1}^{p} \left(\frac{f_{ij}}{f_{i.}} \right) \varphi_j \qquad (1)$$

$$\varphi_j = \beta \sum_{i=1}^{n} \left(\frac{f_{ij}}{f_{.j}} \right) \psi_i \qquad (2)$$

Coefficient β is a positive coefficient that is greater than *1*. We can see that without this coefficient β, each row-point i would be the barycenter (i.e., a weighted center of gravity) of the p column-points, with a system of weights $\left(\frac{f_{ij}}{f_{i.}} \right)$, such that: $\sum_{j=1}^{p} \left(\frac{f_{ij}}{f_{i.}} \right) = 1$. These weights are nothing but the components of the row-profile of element i.

Similarly, each column-point j would be the barycenter of the n row-points, with a system of weights $\left(\frac{f_{ij}}{f_{.j}} \right)$, such that: $\sum_{i=1}^{n} \left(\frac{f_{ij}}{f_{.j}} \right) = 1$.

These weights are nothing but the components of the column-profile of element j [1]. These transition relationships simply involving the original profiles account for the simultaneous representation of rows and columns on the graphical displays (cf. figure 3.3)[2].

Another property of the row-point (and column-point) coordinates on the principal axes is that they are centered, and thus obey these relationships:

$$\sum_{i=1}^{n} f_{i.} \psi_i = 0, \qquad \sum_{j=1}^{p} f_{.j} \varphi_j = 0$$

3.2.3 Validity of the representation

Table 3.4 shows the values of two series of parameters which we have not yet mentioned: the eigenvalues, designated by λ_α (whose values are between *0* and *1* in correspondence analysis)[3].

---

[1] A double-barycentric representation is unfortunately not possible, because the use of a barycenter has a contracting effect. In order for equations (1) and (2) to be possible simultaneously, we therefore need a dilating coefficient β that is greater than *1*.

[2] There is no equivalent property in principal components analysis. Similar relationships do exist, but the weights are not positive, and do not add to one.

[3] In correspondence analysis of an (n, p) contingency table, with $p \leq n$, there are at most p non-zero eigenvalues, *whichever space is involved* (rows or columns) (this is a general property of singular value decomposition; see appendix 1). If the origin of the principal axes is the center of gravity of the set of profile-points, we have at most $(p-1)$ non-zero eigenvalues.

Table 3.4
Eigenvalues and percentages of variance

	EIGEN VALUE	PERCENT.	CUMUL. PERCENT.	
1	.0354	57.04	57.04	**********************************
2	.0131	21.13	78.17	*************
3	.0073	11.76	89.94	*****
4	.0062	10.06	100.00	****

The eigenvalues measure the variances along each principal axis. These eigenvalues are $\lambda_1=0.035$ for the first axis and $\lambda_2=0.013$ for the second axis; the *percentages of variance* (ratio of each eigenvalue to their overall sum t multiplied by *100*), corresponding to these eigenvalues are *57%* and *21%* for the first two axes.

Properties of trace t

Another parameter, the *trace* t (sum of all the eigenvalues) is equal to *0.062*: there are *4* non-zero eigenvalues. Whereas the trace represents the total variance of the configuration of points, the eigenvalues represent the variances corresponding to each axis.

In a correspondence analysis of an (n,p) contingency table, the product of trace t and the total count k is none other than the classical chi-square statistic χ^2 of Karl Pearson, with $(n-1)(p-1)$ degrees of freedom, that is used for testing the independence of a table's rows and columns.

Thus this parameter is calculated with the formula:

$$\chi^2 = kt$$

specifically:

$$\chi^2 = k\sum_{i=1}^{n}\sum_{j=1}^{p}\left(\frac{f_{ij} - f_{i.}f_{.j}}{f_{i.}f_{.j}}\right)^2$$

We have, for table 3.1:

$$kt = 1592 \times 0.062 = 98.7$$

with *52* degrees of freedom [*52 = (14-1)* x *(5-1)*].

The hypothesis of independence is of course rejected: correspondence analysis exists precisely to help us understand why this hypothesis is rejected.

This example shows how inferential statistics and exploratory data analysis complement one another: in the present case, the latter takes over from the former because the nature of the data is so complex that hypothesis testing and model-building are both inappropriate.

Eigenvalues

For a given axis there exists the following relationship between coefficient β of the transition equations and the eigenvalue λ :

$$\beta = \frac{1}{\sqrt{\lambda}}$$

Thus an eigenvalue close to *1* assures a good barycentric representation along the corresponding axis. In such a situation it is possible to interpret the relative positions of the two sets of row-points and column-points[1].

But this is not really the case in our example: if educational-level points were represented as the actual barycenters of the word-points, they would be much closer to the center. This is why the simultaneous representation may seem misleading to the uninitiated analyst.

Percentages of variance

Percentages of variance measure the relative importance of each eigenvalue in the trace. In this case, the plane spanned by the first two principal axes "explains" *78%* of the total variance.

Generally these percentages are a rather conservative measure of the part of information that is being represented. Many examples have been observed where a satisfactory structure of the data was obtained even though these percentages were low. This means that significant patterns or groupings can be present even when there is much noise in the data.

These problems will be addressed again in section 3.2.5 (a comparison with principal components analysis) and in section 3.3.2 (validity of the representation in the case of multiple corespondence analysis).

Other aids to interpretation

There are two series of parameters that are used to help with the interpretation of the results, complementing the information provided by the coordinates of the elements on the principal axes:

---

1 The square root of the largest eigenvalue can also be interpreted as the greatest correlation coefficient that exists between the rows and columns of the table (canonical correlation). This coefficient is calculated in the following fashion: suppose that for each of the *5* categories of educational level there is a corresponding numerical value (5 different values altogether), and that the same is true for the *14* forms (this is called a coding or quantification procedure). For each occurrence there are two numerical values (one for the rows, one for the columns). Therefore we can calculate a correlation coefficient between the two sets of values. The *1592* occurrences can be regrouped into *70* groups (the *70* cells of table 3.1) corresponding to distinct pairs of values. The maximum value that can be attained by this correlation coefficient (which is associated with "optimal coding" of the words and educational levels) is 1. Finding this maximum correlation comes from the work of Hirschfeld (1935).

a) The *contributions* (or absolute contributions), that describe the proportion of variance explained provided by each element (row or column) in building a principal axis.

b) The *squared correlations* (or relative contributions) that measure the quality of representation of each element by the axes.

Table 3.5 shows the values of these parameters for the example presented earlier.

Appendix 1 explains how these parameters are calculated.

Guide to reading table 3.5

- Column Rel. Wt. (Relative weights) designates the marginals of the rows and columns already shown in tables 3.2 and 3.3.

- Column DISTO (Distances from Origin) contains the squares of the distances to the axes' origin, that is, the distances from each profile to the mean profile (or marginal). For example, on table 3.2 the words *finances* and *decision* have profiles that are very different from the marginal.

  These distances are chi-square distances computed from the original data, or, *equivalently*, Euclidean distances computed from the coordinates upon *all* the principal axes.

- The first two COORDINATES are those of the points, whose representation is given on figure 3.1. Note that, to save space and to make the display more legible, some points (such as "Finances", "Economic") far away from the origin have been brought back to the frame of the picture.

- The CONTRIBUTIONS (or absolute contributions), whose column sum equals *100*, show the importance of the different elements in the construction of each axis.

- The SQUARED CORRELATIONS (or relative contributions), whose row sum equals *1*, show the importance of the different axes in explaining the variance of each element.

Table 3.5

Main parameters of correspondence analysis of table 3.1
Rows are active individuals of table 3.1, columns are active variables

			COORDINATES				*CONTRIBUTIONS*				*SQUARED CORREL.*			
ACTIVE COLUMNS	Rel.Wt.	DISTO	1	2	3	4	1	2	3	4	1	2	3	4
No Degree	20.29	.06	.21	.08	-.07	.10	25.1	10.1	14.7	29.9	.68	.10	.08	.14
Elem. Sch.	33.73	.03	.14	-.06	.02	-.08	18.3	8.1	1.5	38.4	.64	.11	.01	.24
Trade Sch.	20.23	.04	-.11	.03	.15	.06	6.8	1.3	59.9	11.9	.31	.02	.57	.10
High Sch.	17.90	.10	-.27	.12	-.08	-.06	38.0	20.1	14.4	9.6	.76	.15	.06	.03
College	7.85	.17	-.23	-.32	-.09	.09	11.9	60.5	9.5	10.3	.31	.59	.05	.05

			COORDINATES				*CONTRIBUTIONS*				*SQUARED CORREL.*			
Words	Rel.Wt.	DISTO	1	2	3	4	1	2	3	4	1	2	3	4
Money	12.12	.03	.12	-.02	-.10	.08	4.5	.4	16.9	13.9	.43	.01	.33	.23
Future	19.97	.04	-.18	.10	.05	-.01	17.6	14.6	7.6	.1	.72	.22	.06	.00
Unemployment	17.78	.05	.21	.07	.00	-.04	22.6	6.8	.0	4.0	.87	.10	.00	.03
Decision	1.38	.28	-.40	-.33	.02	-.07	6.3	11.5	.0	1.0	.58	.40	.00	.02
Difficult	1.70	.07	.25	-.07	-.06	.00	3.0	.6	.8	.0	.88	.06	.05	.00
Economic	3.39	.26	-.35	-.32	-.08	.15	12.0	26.6	3.3	13.0	.48	.40	.03	.09
Selfishness	6.72	.05	-.06	.03	-.18	-.11	.7	.3	29.5	13.6	.07	.01	.66	.26
Occupation	4.96	.11	.14	-.22	.21	-.06	2.6	17.6	30.8	2.7	.16	.41	.40	.03
Finances	1.76	.20	.24	.21	.04	.32	2.8	5.7	.5	29.0	.28	.21	.01	.51
War	1.63	.06	-.22	.07	.10	.02	2.2	.7	2.1	.2	.75	.09	.15	.01
Housing	3.27	.06	.01	-.13	-.09	-.19	.0	4.1	3.5	19.4	.00	.27	.13	.60
Fear	9.99	.05	-.20	.06	.03	-.01	11.7	2.6	1.5	.1	.90	.07	.02	.00
Health	5.84	.02	-.11	.00	-.02	.05	2.1	.0	.4	2.4	.80	.00	.03	.17
Work	9.48	.06	.21	-.11	.05	.02	12.0	8.6	3.0	.6	.75	.20	.04	.01

3.2.4 Active and supplementary variables

Correspondence analysis is used for finding subspaces to represent proximities among profiles. But it can also be used for positioning *supplementary* rows and columns of the data matrix in this subspace.

A nice property of correspondence analysis, inherited from singular value decomposition (see appendix 1), is the identity of non-zero eigenvalues obtained in both spaces. Each pair of axes (Φ_α, Ψ_α) correspond to the same eigenvalue λ_α. From now on, we will drop the subscript α, and focus on a particular axis.

Once the principal axes Φ and Ψ and the eigenvalue λ (and thus β) have been computed, the transition equations (1) and (2) from above can be applied to supplementary rows or columns. Equation (1) is used to calculate, starting with φ, β, for each principal axis, the coordinate ψ_i^+ of a supplementary row i based on the profile $f_{ij}^+ / f_{i.}^+$ (cf. equation 1').

$$\psi_i^+ = \beta \sum_{j=1}^{p} \frac{f_{ij}^+}{f_{i.}^+} \varphi_j \qquad (1')$$

$$\varphi_j^+ = \beta \sum_{i=1}^{n} \frac{f_{ij}^+}{f_{.j}^+} \psi_i \qquad (2')$$

Equation 2 is used in analogous fashion to calculate, starting with ψ, β, the coordinate φ_j^+ of a supplementary column j (cf. equation 2') whose profile is $f_{ij}^+ / f_{.j}^+$. Thus we can illustrate the two dimensional display with supplementary information that did not participate in the construction of the planes. This has very significant consequences in terms of interpreting the results.

The elements or variables used to calculate the two dimensional display are called *active elements* or *active variables*. These elements or variables must be a homogeneous set in order for the distances among individuals or observations to make sense, and therefore the observed graphical proximities to be interpretable. Elements or variables that are projected *a posteriori* on the two dimensional display are the *supplementary* or *illustrative* elements. It is not necessary for these illustrative elements (rows or columns) to constitute a homogeneous set. The computation is executed separately for each one. This dichotomy between active variables and illustrative variables is fundamental from a methodological viewpoint.

Example

The distributions of four words having small overall counts are shown on table 3.6. These words were not part of the preceding analysis.

Table 3.6

Four supplementary (illustrative) rows

Words	No degree	Elem. Sch.	Trade Sch.	High Sch.	Coll ege	Total
Comfort	*2*	*4*	*3*	*1*	*4*	*14*
Disagreement	*2*	*8*	*2*	*5*	*2*	*19*
World	*1*	*5*	*4*	*6*	*3*	*19*
Live	*3*	*3*	*1*	*3*	*4*	*14*

We would like to see where they are situated with respect to the other words represented on the two dimensional display of figure 3.2. Their row-profiles can be positioned in the same 5 dimensional space, and can therefore be projected onto the plane of figure 3.2.

Table 3.7

Three supplementary (illustrative) columns

Word	Age-30	Age-50	Age+50
Money	*59*	*66*	*70*
Future	*115*	*117*	*86*
Unemployment	*79*	*88*	*177*
Decision	*9*	*8*	*5*
Difficult	*2*	*17*	*18*
Economic	*18*	*19*	*17*
Selfishness	*14*	*34*	*61*
Occupation	*21*	*30*	*28*
Finances	*8*	*12*	*8*
War	*7*	*6*	*13*
Housing	*10*	*27*	*17*
Fear	*48*	*59*	*52*
Health	*13*	*29*	*53*
Work	*30*	*63*	*58*
Total	*433*	*575*	*663*

In analogous fashion, table 3.7 contains three supplementary columns (age categories) that were not included in the set of active columns due to the heterogeneous nature of the themes: interpreting the proximities among rows, and thus among words, would have been more difficult. Are two words close because of their distribution with respect to educational level or with respect to age categories? This type of decision is not easy to make if the distances among words are calculated on the basis of both variables simultaneously.[1]

[1] However it is possible to examine the juxtaposition of several contingency tables (cf. chapter 5, section 5.4) in order to obtain a preliminary visualization of a set of characteristics (for example, a set of socio-demographic variables).

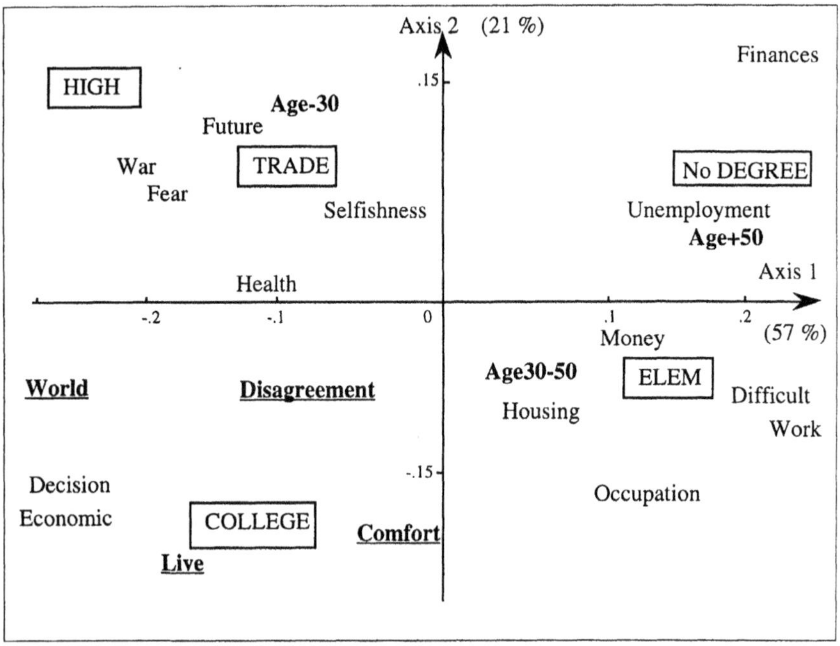

Figure 3.4

Associations among words and Educational Level (*continuation*)
Positioning of illustrative elements (in bold) in plane of figure 3.2

Figure 3.4 shows us that three supplementary words are relatively closely associated with the responses of persons with a higher degree, whereas the fourth, *disagreement*, is less characteristic, being closer to the center that represents the mean profile.

In table 3.8 is found a set of technical parameters that show the special nature of each of these distributions.

It is not surprising that this table does not show *contributions*, since the illustrative elements have a null contribution by definition: they do not participate in the construction of the axes, but are positioned *a posteriori*.

It is good practice to start by using, for active tables, homogeneous data sets that describe proximities from a single point of view. The representation can then be enriched by illustrating it with supplementary information.

On figure 3.4 the three age categories are ordered along the horizontal axis just as the educational levels are: increasing age groups correspond to decreasing educational levels. This is a structural trait of the population under analysis: the younger respondents have more schooling, and this complicates interpretation in terms of causality. One is led to wonder whether the effect of Educational Level on open-ended responses can be separated from the effect of age.

Table 3.8

Parameters of analysis for illustrative elements

Illustrative columns

Column Labels	REL.WT	DISTO	COORDINATES 1	2	3	SQUARED CORREL. 1	2	3
less than 30 years	*27.2*	*.08*	*-.11*	*.06*	*.10*	*.14*	*.04*	*.13*
from 30 to 50 yrs	*36.1*	*.03*	*.02*	*-.05*	*.02*	*.01*	*.09*	*.01*
more than 50 yrs	*41.6*	*.11*	*.18*	*.05*	*-.10*	*.29*	*.02*	*.09*

Illustrative rows

Words	REL.WT	DISTO	COORDINATES 1	2	3	SQUARED CORREL. 1	2	3
Comfort	*.8*	*.64*	*-.21*	*-.70*	*-.07*	*.07*	*.78*	*.01*
Disagreement	*1.1*	*.16*	*-.15*	*-.12*	*-.17*	*.13*	*.09*	*.18*
World	*1.1*	*.31*	*-.52*	*-.14*	*-.08*	*.88*	*.07*	*.02*
Live	*.8*	*.68*	*-.31*	*-.50*	*-.52*	*.14*	*.37*	*.40*

To enhance the interpretation, variables *Age* and *Educational Level* should be cross-tabulated, partitioning the respondents into $3 \times 5 = 15$ columns.

In chapter 5, we shall go beyond the scope of this simple example which is meant to illustrate how the method works.[1]

3.2.5 A comparison with principal components analysis

Both correspondence analysis (CA) and principal components analysis (PCA) are closely related to the general property of rectangular matrices known as *singular value decomposition* (see appendix 1).

Two properties of correspondence analysis are undoubtedly in favour of its use when dealing with contingency tables: on the one hand, the symmetrical role played by rows and columns (there is no *a priori* distinction between variables and observations), on the other, the consistency of the chi-square distance with the underlying multinomial model.

---

[1] An age-educational level cross-tabulation with *9* cells is used in the "actual size" example of chapter 5.

Although it is clear that CA is appropriate for count data or binary data and PCA for real valued measurement, the user of this latter method (much more widespread) may legitimately ask what are the risks of false results when applying it to count or binary data.

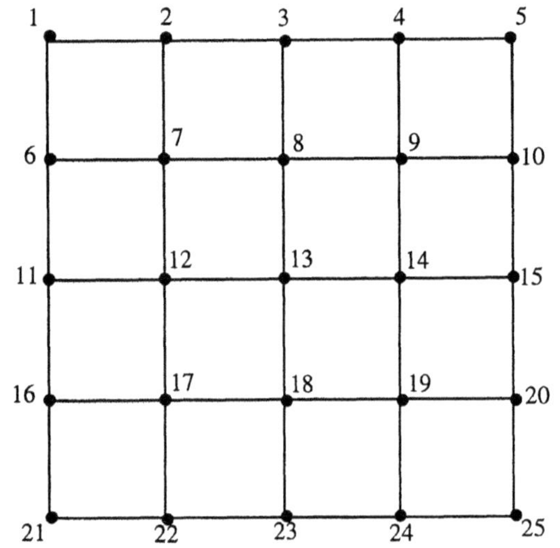

Figure 3.5
Graph G associated with a "chessboard"
(square lattice grid)

Evidently, the availability of a method in various software packages does not justify its systematic use, even in an exploratory phase less demanding from a theoretical standpoint. However, it may prove useful to proceed to some comparisons using readily available data sets. In this respect, it seems natural to calibrate a visualization tool on artificial data sets provided with an *a priori* structure.

We present below a comparison of the two methods applied to a same binary data matrix associated with a "chessboard shaped graph", (figure 3.5).

In this figure, a line (more precisely: an *edge*) drawn between two *vertices* such as 7 and 12, for example, means that the vertices 7 and 12 are *adjacent*.

Table 3.9 contains an example of a numerical coding of the previous *undirected graph*, in which each vertex is supposed to be adjacent to itself. We read in table 3.9, for instance, that vertex 1 is adjacent to vertices 1, 2, 6, that vertex 2 is adjacent to vertices 2, 1, 7, 3, etc. Since the graph is undirected, the pair (7,12) appears twice in table 3.9 (in both row 7 and row 12).

Such tables associated with sparse matrices are not uncommon in textual data. A semantic network between 25 textual words could lead to a table of the same nature ("word 1" has, as semantic neighbours: itself, "word 2" and "word 6", etc.).

Table 3.9
Numerical coding of graph G

Vertex	*Adjacent vertices*				
1	1	2	6		
2	2	1	7	3	
3	3	2	8	4	
4	4	3	9	5	
5	5	4	10		
6	6	1	7	11	
7	7	2	6	8	12
8	8	3	7	9	13
9	9	4	8	10	14
10	10	5	9	15	
11	11	6	12	16	
12	12	7	11	13	17
13	13	8	12	14	18
14	14	9	13	15	19
15	15	10	14	20	
16	16	11	17	21	
17	17	12	16	18	22
18	18	13	17	19	23
19	19	14	18	20	24
20	20	15	19	25	
21	21	16	22		
22	22	21	17	23	
23	23	18	22	24	
24	24	19	23	25	
25	25	24	20		

Table 3.10 shows the symmetric binary sparse matrix associated with the graph. Its general entry (i,j) has value of *1* if the edge (i,j) is mentioned in table 3.9, and the value of *0* otherwise. A specific pattern in the distribution of the "*1*" is already apparent in table 3.10, due to the particular sequential numbering of the vertices.

A random permutation of the rows (or of the columns) would suffice to hide this apparent pattern without modifying the underlying structure.

After such a permutation, visualization tools enabling to uncover the original graph structure are needed all the more.

Table 3.10

Sparse binary matrix M associated with graph G

	1	2	3	4	5	6	7	8	9	10	11	12	13	14	15	16	17	18	19	20	21	22	23	24	25
r01	1	1	0	0	0	1	0	0	0	0	0	0	0	0	0	0	0	0	0	0	0	0	0	0	0
r02	1	1	1	0	0	0	1	0	0	0	0	0	0	0	0	0	0	0	0	0	0	0	0	0	0
r03	0	1	1	1	0	0	0	1	0	0	0	0	0	0	0	0	0	0	0	0	0	0	0	0	0
r04	0	0	1	1	1	0	0	0	1	0	0	0	0	0	0	0	0	0	0	0	0	0	0	0	0
r05	0	0	0	1	1	0	0	0	0	1	0	0	0	0	0	0	0	0	0	0	0	0	0	0	0
r06	1	0	0	0	0	1	1	0	0	0	1	0	0	0	0	0	0	0	0	0	0	0	0	0	0
r07	0	1	0	0	0	1	1	1	0	0	0	1	0	0	0	0	0	0	0	0	0	0	0	0	0
r08	0	0	1	0	0	0	1	1	1	0	0	0	1	0	0	0	0	0	0	0	0	0	0	0	0
r09	0	0	0	1	0	0	0	1	1	1	0	0	0	1	0	0	0	0	0	0	0	0	0	0	0
r10	0	0	0	0	1	0	0	0	1	1	0	0	0	0	1	0	0	0	0	0	0	0	0	0	0
r11	0	0	0	0	0	1	0	0	0	0	1	1	0	0	0	1	0	0	0	0	0	0	0	0	0
r12	0	0	0	0	0	0	1	0	0	0	1	1	1	0	0	0	1	0	0	0	0	0	0	0	0
r13	0	0	0	0	0	0	0	1	0	0	0	1	1	1	0	0	0	1	0	0	0	0	0	0	0
r14	0	0	0	0	0	0	0	0	1	0	0	0	1	1	1	0	0	0	1	0	0	0	0	0	0
r15	0	0	0	0	0	0	0	0	0	1	0	0	0	1	1	0	0	0	0	1	0	0	0	0	0
r16	0	0	0	0	0	0	0	0	0	0	1	0	0	0	0	1	1	0	0	0	1	0	0	0	0
r17	0	0	0	0	0	0	0	0	0	0	0	1	0	0	0	1	1	1	0	0	0	1	0	0	0
r18	0	0	0	0	0	0	0	0	0	0	0	0	1	0	0	0	1	1	1	0	0	0	1	0	0
r19	0	0	0	0	0	0	0	0	0	0	0	0	0	1	0	0	0	1	1	1	0	0	0	1	0
r20	0	0	0	0	0	0	0	0	0	0	0	0	0	0	1	0	0	0	1	1	0	0	0	0	1
r21	0	0	0	0	0	0	0	0	0	0	0	0	0	0	0	1	0	0	0	0	1	1	0	0	0
r22	0	0	0	0	0	0	0	0	0	0	0	0	0	0	0	0	1	0	0	0	1	1	1	0	0
r23	0	0	0	0	0	0	0	0	0	0	0	0	0	0	0	0	0	1	0	0	0	1	1	1	0
r24	0	0	0	0	0	0	0	0	0	0	0	0	0	0	0	0	0	0	1	0	0	0	1	1	1
r25	0	0	0	0	0	0	0	0	0	0	0	0	0	0	0	0	0	0	0	1	0	0	0	1	1

Principal components analysis of matrix M

In a first step, principal components analysis is applied to data matrix **M**. Such an analysis can be performed using either the covariance matrix or the correlation matrix. The numerical results appear to be similar in both cases, the obtained visualizations being almost identical.

Thus the analysis involving the correlation matrix is presented here. Figure 3.6 shows a visualization of the locations of the *25* vertices in the plane spanned by the first two principal axes.

These axes correspond to two identical eigenvalues ($\lambda_1 = \lambda_2 = 3.98$), explaining together *31.86* % of the total variance.

The vertices adjacent in the original graph have been joined by an edge to highlight the initial structure.

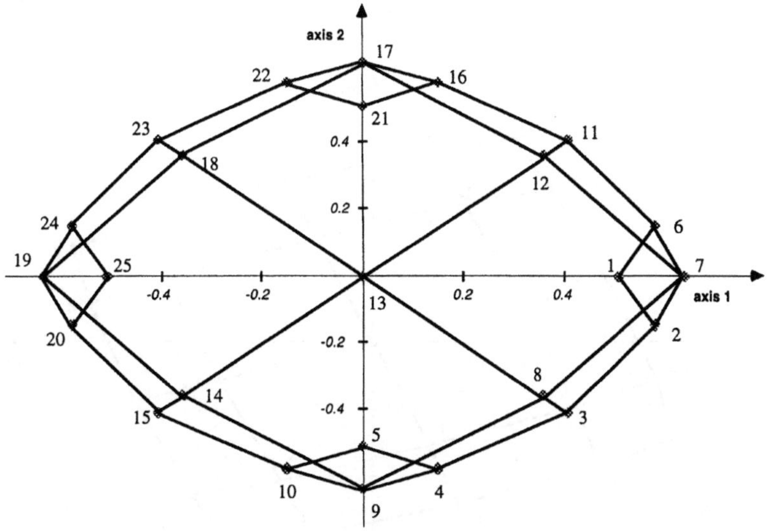

Figure 3.6
Visualization of graph G through principal components analysis (plane spanned by the first two principal axes)

The symmetry with respect to vertex number 13 is reconstituted. The relative locations of the vertices vis-à-vis their neighbours is generally taken into account by the display, with the exception of the four vertices corresponding to the corners of the rectangle (vertices 1, 5, 21, 25) that are folded back toward the center. The changes in the lengths of some edges are noticeable. They are characterized by a dilation of the four most central cycles of the graph.

Correspondence analysis of matrix M

Correspondence analysis is then applied to the same data matrix **M**. Figure 3.7 shows a visualization of the locations of the *25* vertices in the plane spanned by the first two principal axes. These axes also correspond to two identical eigenvalues ($\lambda_1 = \lambda_2 = 0.814$), explaining together *32.24* % of the total variance.

Although the graph in figure 3.5 is somewhat conventional (it can be drawn in several possible ways), the display in figure 3.7 satisfactorily reconstitutes both the relative positions of the vertices and an acceptable order of magnitude for the lengths of the various edges. This ability of CA to produce legible maps out of such data matrices can be extended to binary matrices describing various planar graphs[1].

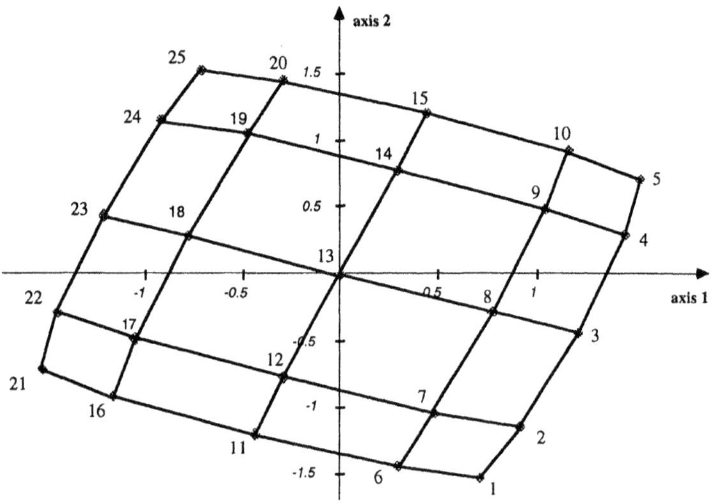

Figure 3.7
Visualization of graph G through correspondence analysis (plane spanned by the first two principal axes)

We note that the percentage of explained variance (*32.24 %*) is relatively modest if confronted with the quality of the reconstitution of the original structure in the corresponding plane. In CA (and in PCA as well), this phenomenon often occurs when dealing with binary data (see also section 3.3.2 below). In this context, the percentages of variance explained by principal axes always give a pessimistic view of the extracted information.

In conclusion, besides the theoretical arguments previously mentioned and developed in appendix 1, a variety of empirical evidence favours the use of CA in the case of binary and count data. The superiority of correspondence

---

[1] It can be shown, for example, that the coordinates of the vertices of a regular graph, on the first principal axis of a CA of its associated binary matrix, have a minimal *Geary contiguity ratio* on the graph. This result implies that, on the average, the coordinates of adjacent vertices are as much as possible similar. See Lebart (1984), Burtschy and Lebart (1991).

analysis in some obvious cases does not necessarily mean that analyses making use of prinçipal components analysis or factor analysis for count data are obsolete. Each exploratory tool provides a specific point of view over the data set. In an exploratory context, utilizing a versatile tool such as PCA, be it sub-optimal, might be more important and productive than an awkward handling of another technique.

3.3 MULTIPLE CORRESPONDENCE ANALYSIS

Multiple correspondence analysis is used to obtain descriptions of large binary matrices, such as data from socio-economic surveys. The rows of these matrices are generally individuals or observations (of which there can be up to several thousand). The columns are categories of nominal variables, most often response categories of questions. The principles of this method are due to Guttman (1941), Burt (1950) and Hayashi (1956). Multiple correspondence analysis is in reality a simple extension of the sphere of application of correspondence analysis, with its own procedures for computation and interpretation of results.

This extension is based upon the following property. Suppose k individuals are asked two questions with n and p response categories respectively, and that for each question the various categories are mutually exclusive (i.e., only one response is allowed).

Then it is equivalent to analyze with correspondence analysis the contingency table (n,p) that cross-tabulates the two questions, or to analyze the binary matrix $\mathbf{Z}$ *with k rows and (n+p) columns that describes the responses.*[1]

Analyzing $\mathbf{Z}$ is computationally more expensive but more interesting because it can be easily generalized to the case of more than two questions.

Thus table 3.11 represents the case of three questions having 4, 3 and 4 response categories respectively.

The complete disjunctive matrix $\mathbf{Z}$ consists of three blocks. The responses of the *12* individuals (rows) can thus be coded into a reduced matrix $\mathbf{R}$, (*12* rows and *3* columns) that contains the numbers of the appropriate category. In practice, the computational procedures use the reduced matrix.

The third matrix $\mathbf{B}$, the product of the disjunctive matrix $\mathbf{Z}$ by its transpose, is shown in table 3.12.

$$\mathbf{B} = \mathbf{Z'Z}$$

[1] In this case each row of $\mathbf{Z}$ has two "*1*"s in the columns that correspond to chosen answers, and $n+p-2$ "*0*"s.

This matrix is symmetric (it is often called *Burt's table* or *multiple correspondence table*). It has nine blocks.

Table 3.11
Matrices R (reduced coding), Z (complete disjunctive)

R =

2	3	4
2	1	3
3	1	2
4	2	4
1	2	3
2	2	3
3	1	1
1	1	1
4	1	2
2	2	3
3	2	2
4	1	4

Z =

0	1	0	0	0	0	1	0	0	0	1
0	1	0	0	1	0	0	0	0	1	0
0	0	1	0	1	0	0	0	1	0	0
0	0	0	1	0	1	0	0	0	0	1
1	0	0	0	0	1	0	0	0	1	0
0	1	0	0	0	1	0	0	0	1	0
0	0	1	0	1	0	0	1	0	0	0
1	0	0	0	1	0	0	1	0	0	0
0	0	0	1	1	0	0	0	1	0	0
0	1	0	0	0	1	0	0	0	1	0
0	0	1	0	0	1	0	0	1	0	0
0	0	0	1	1	0	0	0	0	0	1

The diagonal blocks are diagonal matrices whose diagonal elements are the counts of responses corresponding to each category.

The 6 remaining blocks are the three contingency tables, that cross-tabulate the three questions pairwise, and their transposes.

Table 3.12
B=Z'Z Matrices (Burt's table)

B = Z'Z =

2	0	0	0	1	1	0	1	1	1	1
0	4	0	0	1	2	1	0	0	3	1
0	0	3	0	2	1	0	1	2	0	0
0	0	0	3	2	1	0	0	1	0	2
1	1	2	2	6	0	0	2	2	1	1
1	2	1	1	0	5	0	0	1	3	1
0	1	0	0	0	0	1	0	0	0	1
1	0	1	0	2	0	0	2	0	0	0
1	0	2	1	2	1	0	0	3	0	0
1	3	0	0	1	3	0	0	0	4	0
1	1	0	2	1	1	1	0	0	0	3

Correspondence analysis of matrix **Z** yields the same normalized factor axes as does that of matrix **B**, and very similar proximity maps. However the scaling is different, since the eigenvalues are different: eigenvalues of the analysis of **Z** are the square roots of the eigenvalues of the analysis of **Z'Z**.

The case of two questions

In the case of two questions, matrix **R** only has *2* columns, matrix **Z** is formed of 2 blocks, $\mathbf{Z} = (\mathbf{Z_1},\mathbf{Z_2})$ and **B** has *4*.

B can be written: $$\mathbf{B} = \begin{pmatrix} \mathbf{Z}_1' \mathbf{Z}_1 & \mathbf{Z}_1' \mathbf{Z}_2 \\ \mathbf{Z}_2' \mathbf{Z}_1 & \mathbf{Z}_2' \mathbf{Z}_2 \end{pmatrix}$$

The two preceding analyses also give the same representation as the correspondence analysis of the small contingency table $\mathbf{C} = \mathbf{Z}_1' \mathbf{Z}_2$ cross-tabulating the two questions (matrix **C** is a non-diagonal block of **B**).

Note that in this case, the two questions that make up the *rows and columns* of contingency table **C** correspond only to the *columns* of matrix **Z**. Thus the simultaneous representation of the rows and columns is not merely a graphic artifact: it finds its justification in this presentation of correspondence analysis.

3.3.1 Basic structure of a survey sample

The preceding example showed how an *Educational Level* effect could conceal an *Age* effect (in our sample the least educated respondents are generally the oldest). Given the structure of populations under study these effects are not independent from gender and living standard.

This observation gives rise to the concept of first describing the network of interrelations among all of the basic characteristics of the respondents and then overlaying the other themes of the survey as illustrative or descriptive elements.

The following example illustrates how multiple correspondence analysis is used for analyzing survey data.

Again, to simplify the presentation of results, the actual scale of the application is reduced: table **Z** has *144* rows and *24* active columns that represent the response categories for *5* questions (table 3.13). Table **Z+** has an additional *23* illustrative questions that show the presence or absence of words extracted from an open-ended question asked among *144* respondents. This question, based on work of C. Baudelot (1988) is worded:

"What does it mean to you to be successful in life ?".

(The corresponding element z_{ij}^{+} of **Z+** is *1* if individual *i* used word *j* for the previously mentioned open-ended response and *0* otherwise. Thus Z+ has

50 columns). In an actual full-size analysis it is not unusual to position several hundred categories as illustrative elements. The real power of this method resides in its ability to systematically sort and filter information on a large scale.

The pedagogical paradox mentioned in paragraph 3.1 still holds: here again there exists a conflict between the size and the relevance of the example.

Table 3.13

List of active and illustrative questions

I - *5* active questions (*15* Categories)

	number of categories
1 . gender (Sex)	*2*
2 . level of Diploma	*4*
3 . marital status	*4*
4 . do you have children ?	*2*
5 . age categories	*3*

II - *23* supplementary variables

(Presence or absence of words in response to open-ended question)

	number of categories		number of categories
money	*2*	husband	*2*
achieve	*2*	profession	*2*
well	*2*	my	*2*
good	*2*	professional	*2*
happiness	*2*	succeed	*2*
job	*2*	success	*2*
balance	*2*	health	*2*
relatives	*2*	feel	*2*
family	*2*	employment	*2*
start	*2*	career	*2*
me	*2*	work	*2*
mine	*2*		

How to read table 3.14

Table 3.14 is the lower half of the Burt's table (which is symmetrical) relative to the *5* active questions. In this table are found all the contingency tables (*10* altogether) that cross-tabulate the *5* active questions pairwise.

Along the diagonal are found the questions cross-tabulated with themselves, and thus the frequencies for each category: thus we read that there are *70* males and *74* females in the sample. We also read that there are *78* single persons (first term of the third diagonal block); among them, there are *42* males and *36* females (first terms of the same row).

Table 3.14

Pairwise cross-tabulations of five active questions (Burt's Table B)

	male	fem.	Elem	Trad.	HS	Coll	Sngl	Marr	Divo	Widw	Chld	Chdn	A–30	A–50	A+50
male	70	0													
fem.	0	74													
Elem	15	12	27	0	0	0									
Trad	12	24	0	36	0	0									
HS	18	25	0	0	43	0									
Coll	25	13	0	0	0	38									
Sngl	42	36	10	19	24	25	78	0	0	0					
Marr	27	28	12	13	18	12	0	55	0	0					
Divo	1	6	2	3	1	1	0	0	7	0					
Widw	0	4	3	1	0	0	0	0	0	4					
Chld	23	29	15	17	15	5	2	42	5	3	52	0			
Chdn	47	45	12	19	28	33	76	13	2	1	0	92			
A–30	39	35	8	16	23	27	65	8	1	0	3	71	74	0	0
A–50	26	23	12	10	17	10	9	35	5	0	34	15	0	49	0
A+50	5	16	7	10	3	1	4	12	1	4	15	6	0	0	21
	male	fem.	Elem	Trad.	HS	Coll	Sngl	Marr	Divo	Widw	Chld	Chdn	A–30	A–50	A+50

Table 3.15
Multiple correspondence analysis parameters for active elements

CATEGORIES			COORDINATES				CONTRIBUTIONS				SQUARED CORREL.			
IDEN – LABEL	REL. WT	DISTO	1	2	3	4	1	2	3	4	1	2	3	4
1 . **Gender**														
male – male	9.72	1.06	–.22	–.49	–.58	–.14	.9	7.3	13.2	.9	.04	.23	.32	.02
fem. – female	10.28	.95	.20	.46	.55	.14	.8	6.9	12.5	.9	.04	.23	.32	.02
			cumulative contribution =				1.7	14.3	25.7	1.8				
3 . **Educational level**														
Elem – None	3.75	4.33	.68	.31	–1.19	–.40	3.3	1.1	21.5	2.8	.11	.02	.32	.04
Trad – Trade School	5.00	3.00	.29	.63	.79	–.40	.8	6.2	12.7	3.9	.03	.13	.21	.05
HS – High School	5.97	2.35	–.07	–.37	.54	1.09	.1	2.6	7.2	33.2	.00	.06	.13	.50
Coll – College	5.28	2.79	–.68	–.39	–.52	–.56	4.6	2.5	5.8	7.8	.16	.05	.10	.11
			cumulative contribution =				8.8	12.4	47.2	47.7				
7 . **Marital Status**														
Sngl – Single	10.83	.85	–.80	.18	.02	.02	13.4	1.1	.0	.0	.76	.04	.00	.00
Marr – Married	7.64	1.62	.89	–.55	–.10	.32	11.6	7.3	.3	3.7	.49	.19	.01	.06
Divo – Divorced	.97	19.57	1.01	.14	1.82	–3.09	1.9	.1	13.2	43.9	.05	.00	.17	.49
Widw – Widowed	.56	35.00	1.65	3.82	–2.27	.53	2.9	25.5	11.6	.7	.08	.42	.15	.01
			cumulative contribution =				29.8	33.9	25.1	48.4				
8 . **Presence of children**														
Chld – yes	7.22	1.77	1.19	–.21	.00	.01	19.5	1.0	.0	.0	.79	.03	.00	.00
Chldn– no	12.78	.57	–.67	.12	.00	–.00	11.0	.6	.0	.0	.79	.03	.00	.00
			cumulative contribution =				30.6	1.6	.0	.0				
9 . **Age** (3 categories)														
A–30 – Less than 30 years	10.28	.95	–.84	.13	.04	.02	13.9	.6	.1	.0	.74	.02	.00	.00
A–50 – Between 30 and 50 y.	6.81	1.94	.78	–.86	.10	–.16	8.1	15.8	.3	.8	.32	.38	.01	.01
A 50 – More than 50 years	2.92	5.86	1.13	1.53	–.37	.29	7.1	21.5	1.6	1.2	.22	.40	.02	.01
			cumulative contribution =				29.1	37.8	2.0	2.0				

How to read table 3.15

The rules for reading table 3.15 are almost completely identical to those for table 3.5 above. Only the computations for the cumulative contributions for the categories of each question are new. Their interpretation is straightforward.

How to read figure 3.8

The structure of the *configuration* of the active categories is described by the two dimensional display in figure 3.8, thus providing a summary of the *10* cross-tabulations. In order to facilitate reading the display, polygonal lines have been drawn connecting the ordered items of variables *Age* and *Educational level.*

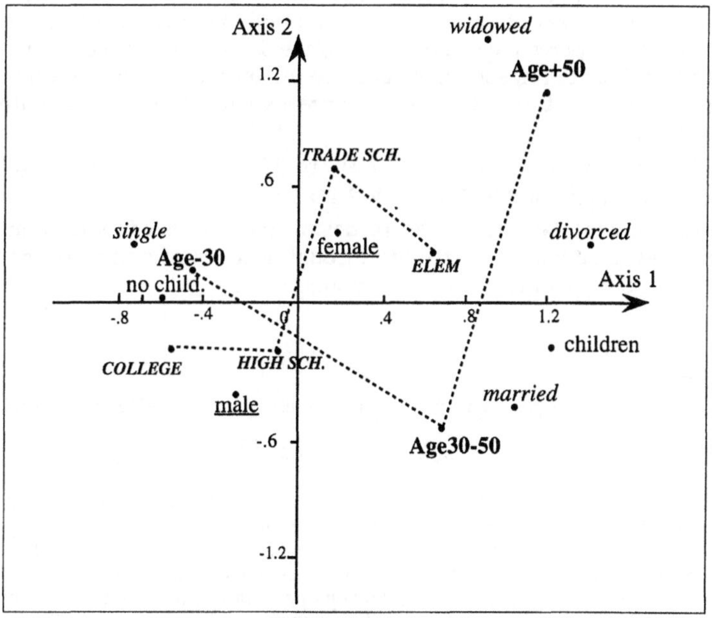

Figure 3.8
Basic structure of sample

Despite the condensed nature of the variables, the connections among *Educational Level, Age, Gender and Marital Status* are more complex than it would seem. A *life cycle* effect and a *generation* effect are hidden behind variables with two categories such as *Gender*: In this study the category "female" is associated with a relatively large number of older people without a degree. This phenomenon has to be taken into consideration every time a possible "effect of this variable" on an opinion or an attitude is interpreted.

However figure 3.8 is only a planar approximation of a set of proximities that cannot always be well represented in a plane. Thus there are some distortions among the proximities shown. In fact, figure 3.8 is an intermediate step rather than a final result.

The main operation is the projection of supplementary variables onto this two dimensional summary of the basic structure.

Before examining the important phases of projection of supplementary elements, let us say a few words about the validity of the representation.

3.3.2 Validity of the representation

In the case of multiple correspondence analysis it is not recommended to use the percentages of variance to make a judgment as to the relevance of the axes: these percentages do not have the same significance when we are dealing with a contingency table because the binary coding introduces *noise* that reduces the proportion of variance explained associated with each eigenvalue.[1]

The first two percentages (rounded) are *26%* and *16%* respectively, for eigenvalues of *0.51* and *0.31* (cf. table 3.16).

In practice the stability of the principal axes is tested with simulation techniques (assessment of stability through a series of perturbations of the data matrix) or with resampling techniques[2].

---

[1] Benzécri (1979) proposes an equation for computing corrected eigenvalue as follows:

$$\rho(\lambda) = \left(\frac{s}{s-1}\right)^2 \left(\lambda - \frac{1}{s}\right)^2$$

where s represents the number of active questions (categorical variables) and λ the eigenvalue resulting from the correspondence analysis of the complete disjunctive table. In addition, when $s=2$ the result is nothing but the eigenvalue of the correspondence analysis of the actual contingency table that cross-tabulates the two questions. Greenacre and Blasius (1994) propose an iterative modification of Burt's table that leads to very similar maps with percentages of variance that fall somewhere between the original and the corrected percentages (under the name of Joint Correspondence Analysis).

[2] In the context of multiple correspondence analysis, the principles of the most frequently used method of resampling (the *Bootstrap* method) are relatively simple: various *replicates* of the original data tables are obtained by randomly drawing the rows (i.e., respondents) with replacement. The columns of the replicated Burt's tables are then projected as supplementary elements (see section 3.3.3. below) onto the original visualization space. The set of replicates of each variable (i.e. each category) allows the user to draw an approximate confidence area around the location of the point representing this category in the original display (see: Greenacre, 1984; Lebart *et al.*, 1995).

Table 3.16
Eigenvalues and percentages of variance

Number	Eigen value	Percentage	Cumulative Percentage	
1	*.51*	*25.95*	*25.95*	*******************
2	*.31*	*15.89*	*41.83*	************
3	*.24*	*12.27*	*54.11*	**********
4	*.21*	*10.59*	*64.69*	********
5	*.19*	*9.88*	*74.58*	******
6	*.16*	*8.03*	*82.61*	*****
7	*.12*	*6.36*	*88.97*	****
8	*.11*	*5.61*	*94.58*	***
9	*.06*	*3.27*	*97.85*	**
10	*.04*	*2.15*	*100.00*	*

Here the small sample size (*144* individuals) makes it impossible to proceed to refined inferences based upon the observed proximities.

Table 3.17
Factor coordinates and contributions for supplementary words

Words	Rel.Wt.	DISTO	COORDINATES			SQUARED CORREL.		
			F1	F2	F3	F1	F2	F3
money	.009	7.68	.23	.17	.39	.007	.004	.020
achieve	.003	43.28	-.50	-.36	-.25	.006	.003	.001
well	.013	6.67	.32	-.34	.15	.016	.017	.004
good	.004	30.55	-.70	.84	.77	.016	.023	.020
happiness	.003	32.85	.78	.03	.47	.019	.000	.007
job	.005	25.45	-.57	.36	.23	.013	.005	.002
balance	.004	13.49	.52	-.46	.58	.020	.016	.025
relatives	.005	14.10	.10	-.24	.22	.001	.004	.004
family	.009	8.22	.17	-.15	.03	.004	.003	.000
start	.002	61.00	-.40	.01	.53	.003	.000	.005
my	.005	17.09	-.89	-.23	-.23	.047	.003	.003
me	.002	34.56	-.47	-.18	-.50	.006	.001	.007
mine	.002	39.23	-.54	-.30	.03	.007	.002	.000
husband	.003	30.40	-.30	.70	1.13	.003	.016	.042
profession	.005	11.97	-.31	.91	.24	.008	.070	.005
professional	.006	13.13	.58	-.55	.70	.026	.023	.038
succeed	.024	2.41	-.20	-.14	.06	.016	.009	.002
success	.004	14.54	.52	-.36	.57	.019	.009	.022
health	.004	22.75	1.30	-.09	-.84	.075	.000	.031
feel	.003	22.88	-.74	-.08	-.17	.024	.000	.001
employment	.007	16.72	-.74	.42	.36	.033	.010	.008
career	.014	5.07	.37	-.06	.30	.027	.001	.018
work	.004	41.29	.75	-.29	-.73	.014	.002	.013

3.3.3 Positioning of supplementary variables

The simultaneous representation of the rows and columns associated with correspondence analysis has not been used here: only the columns of matrix **Z** are shown on figure 3.8. The *144* row-points have not been displayed for reasons of clarity, and also because the corresponding individuals are anonymous; only their characteristics are of interest. However other available information about the respondents can be "projected" as illustrative elements.

The fact that matrix **Z** only has "*0*"s and "*1*"s makes the interpretation of the relative positions of the supplementary columns particularly simple.

Transition equation (2) above shows us that the coordinate φ_j of a column j (illustrative response) is obtained by multiplying by β the simple *arithmetic mean* of the coordinates ψ_i of the individuals that chose response j. This is how each illustrative category is positioned.

How to read figure 3.9

On figure 3.9 is shown the positioning of the words whose labels and coordinates are displayed in table 3.17. These response elements are projected as illustrative variables on figure 3.8 (the categories of the active variables are not shown on figure 3.9 in order to keep it legible).

The existing associations among word usages and respondent characteristics are immediately visible in a context that takes into account the inter-relationships between these characteristics. The usual perusal of cross-tabulations assumes that "any one variable might be hiding another". It is mostly redundant when the various characteristics are highly associated with one another.

The present system of projecting supplementary variables thus saves time and avoids errors of interpretation that might occur when sequentially inspecting a series of contingency tables.

The series of illustrative variables displayed on figure 3.9 is constituted by *23* words (table 3.13) used by the same individuals in their responses to the open-ended question: *"What does it mean to you to be successful in life?"* These *23* words were selected because of their away-from-the-center, thus characteristic positions. In a real-life application hundreds of words are projected and sorted in this fashion.

For example the possessive pronouns (*my, mine*) are found in a zone where the respondents are young and well educated. They are seen along with verbs (*achieve, arrive, succeed*). Only the verb *work* characterizes older people, who also tend to express themselves much less in the first person.

All of these proximities suggest a series of conjectures to be verified on the initial data and on a larger sample.

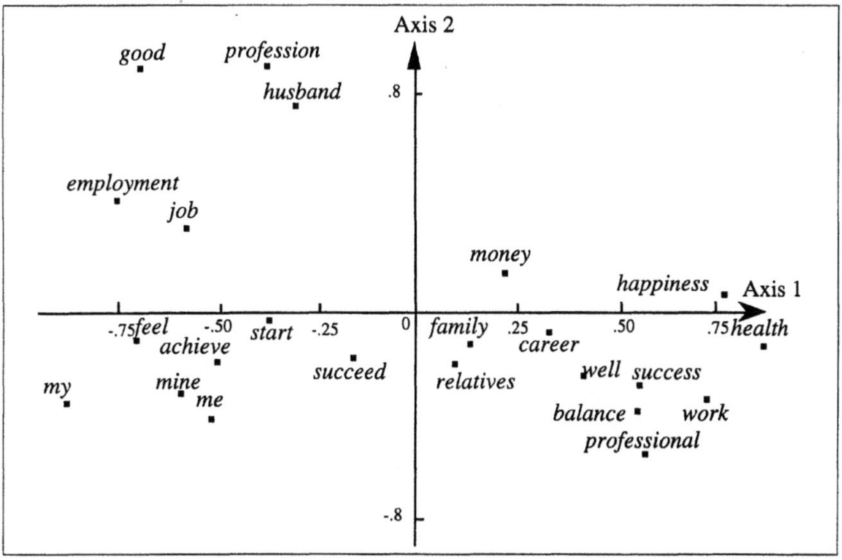

Figure 3.9
Positioning of words of table 3.17 on plane of figure 3.8

Summary

The steps taken for this application of multiple correspondence analysis can be summarized as follows:

- establishing a population typology according to a point of view characterized by a homogeneous set of active variables (in this case: the basic characteristics of the respondents).
- identifying subgroups of this population based on all other available and relevant information. In fact, this method makes it possible to automatically select supplementary variables that have "significant" coordinates on the factor axes, suggesting systematic explorations with multiple cross-tabulations of variables.

Thus simple and multiple correspondence analysis are used to obtain provocative but fragile images of contingency tables and binary coding tables. These images must therefore be handled with caution!

The concurrent use of clustering methods presented in the next chapter allows the analyst to proceed to more secure interpretations and more robust conclusions.

4

CLUSTER ANALYSIS OF WORDS AND TEXTS

Clustering techniques constitute a second family of data analysis techniques in addition to principal axes methods. These methods are used for representing proximities among the elements of a lexical table (rows or columns) through groupings or clusters.

Clustering techniques are divided into two major families:

- Methods of *hierarchical clustering*, which allow us to obtain a hierarchy of groups partially nested in one another, starting with a set of elements that are characterized by variables (or elements whose pairwise distances are known).
- *Partitioning* methods, or direct clustering methods, that produce simple segments or partitions of the population under analysis without the intermediate step of a hierarchical cluster analysis. These methods are better adapted to very large data sets (several thousand individuals).

The methods of both families can be combined into a mixed approach that is discussed in section 4.3.

In practice the results obtained with clustering methods have revealed themselves to be essential complements to the results provided by correspondence analysis, described in the previous chapter.

In analyzing lexical tables, correspondence analysis uncovers major structural traits that apply simultaneously to the two sets (rows and columns) being analyzed. However, when the number of elements is large, it becomes difficult in practice to understand their relative positions by simply looking at graphical displays. The displays are harder to read due to the large number of elements. The same holds true if the text being analyzed is long, even if the number of text parts is not that large. The number of words (or types, or graphical forms) increases rapidly even when the analysis is restricted to units occurring with a frequency greater than a given threshold.

A second reason that motivates a combined use of principal axes methods and clustering has to do with enriching the representations from a multidimensional viewpoint. The groupings obtained are based upon distances calculated *in the whole space*, and not only with the first principal coordinates. Therefore they are able to correct some of the distortions that are inherent in projections into a lower dimensional space.

The final reason for this complementarity is a practical one: it is easier to have the computer describe clusters rather than a continuous space, even if the dimensionality of this space has been drastically reduced. We will show the importance of this "dissection" of space through an example.

This chapter is limited to the description and practice of clustering methods that use the same metric as correspondence analysis, the chi-square metric defined in the preceding chapter, so as to assure compatibility of results.

After a brief review of the methods themselves (4.1) cluster analysis will be used to describe in sequence the proximities on the words and the parts of a corpus (4.2). Then we shall show its role in "sample survey" type data sets (4.3). A very brief technical reminder about cluster analysis is presented in appendix 2.

4.1 REVIEW OF HIERARCHICAL CLUSTER ANALYSIS

Like correspondence analysis, hierarchical cluster analysis[1] is applied to row and column matrices such as the lexical tables described in the preceding chapters. Either the entire set of columns (usually words and sometimes text parts) or the entire set of rows (most often the different text parts) is clustered.

The basic principle of hierarchical agglomeration is very simple. We start with a set of *n* elements (which we shall refer to here as basic elements or *terminal elements*), each of which has a weight, and among which distances have been calculated (then there are $n(n-1)/2$ distances among the possible pairs). First the two elements that are closest to one another are agglomerated. This agglomerated pair now constitutes a *new element* for which both a weight and distances to each of the elements still to be clustered can be calculated. After this step, there remain $n-1$ elements to be clustered. Once again the two closest elements are agglomerated, and this process is repeated $n-1$ times until the set of elements is completely used up.

---

[1] The first book dealing with hierarchical cluster analysis is probably: Sokal and Sneath (1963). Other general manuals are: Hartigan (1975), Gordon (1981), Kaufman and Rousseuw (1990). Interesting critical reviews are given by Cormack (1971), Gordon (1987), Bock (1994).

The last, $(n\text{-}1)^{th}$ operation groups together the whole set of elements into one large cluster, which coincides with the original data set[1] .

Each of the groupings created at each step by following this method is called a *node*. The set of terminal elements corresponding to a node is called a *cluster*. The principle of this method is *agglomerative* clustering, the most prevalent family of methods. There also exist *divisive* methods that operate through successive dichotomizations of the entire population[2].

4.1.1 The dendrogram

The cluster analysis thus obtained can be represented in several different ways. The representation as a *hierarchical tree* or *dendrogram* is probably the most revealing one (see appendix 2).

The grouping obtained at each step of the hierarchical clustering algorithm bring together elements that are the closest to one another. They constitutes a *node* of the hierarchy, and their distance is the *index* attached to the node. The more the grouping procedure progresses, (the closer we come to the top of the tree), the greater the number of already agglomerated elements is, and the greater the minimal distance between the clusters that remain to be agglomerated.

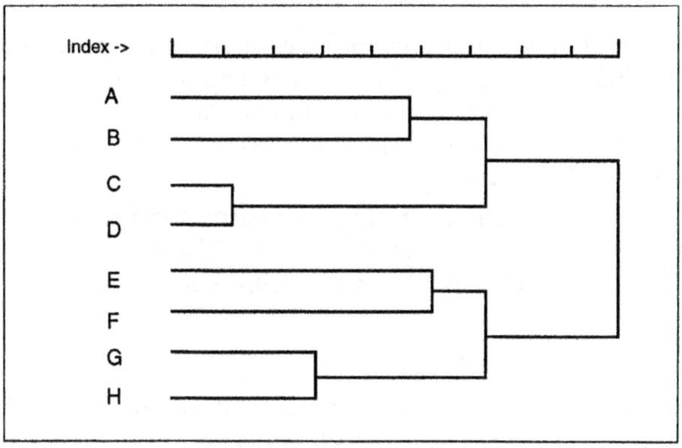

Figure 4.1
Dendrogram representing a cluster analysis on a set of eight elements

1 In practice there are a number of ways of proceeding that fit this definition, which explains the large variety of clustering methods. As indicated above, we only use one of these variants, based on applying the chi-square distance and maximizing the second-order moment of a partition, sometimes called the generalized Ward's criterion (see appendix 2).

2 One method of this type was recommended by Reinert (1983) in the textual field for clustering *context units*.

On figure 4.1 we see that elements C and D, being very close, are agglomerated at the first iteration, and that the two blocks constituted by [A,B,C,D] on the one hand, and [E,F,G,H] on the other hand, which become agglomerated last, probably constitute the two clusters of a good partition of the population (i.e. of the eight terminal elements) with respect to the distance used.

The representation as a dendrogram of a hierarchical cluster analysis shows very concretely the fact that the groups created in the course of a cluster analysis constitute an *indexed hierarchy* of clusters that are partially nested in one another.

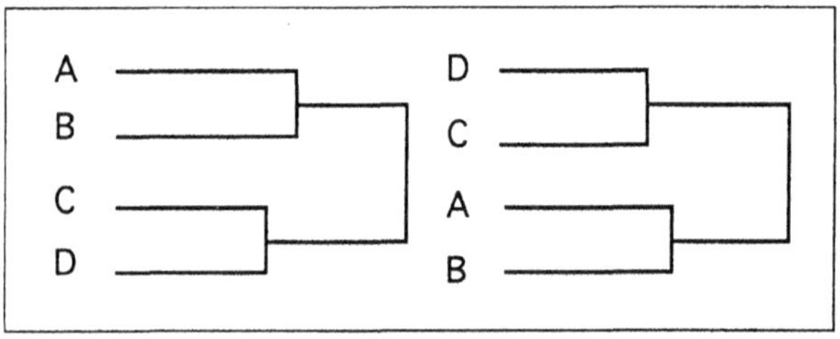

Figure 4.2
Equivalent dendrograms

Interpretation of this hierarchy includes analyzing the distances among elements or clusters that are part of the same node. In fact, for a cluster such as [A,B,C,D], the dendrogram is not unique because each permutation of the first and last element within each node yields an equivalent representation of the hierarchy of clusters.

Thus the two representations of figure 4.2 correspond to the same clustering of the four elements of a hierarchy that includes, besides the terminal elements, the three clusters [C,D], [A,B], [A,B,C,D].

4.1.2 Cutting the dendrogram

If the number of elements to cluster is large, it becomes difficult to examine the complete representation of the classification tree. A practical solution consists in defining a *cutting level* of the tree (which corresponds exactly to an interval between two iterations of the process of agglomerative clustering). Such a cutting makes it possible to examine a set of clusters that summarizes the grouping levels that are higher up in the hierarchy.

Alternatively, we can define where to cut the dendrogram by determining in advance the overall number of clusters desired or equivalently the number of higher nodes to be kept for the cluster set (the number of groups kept is equal to the number of nodes plus one).

The cutting done on the dendrogram shown in figure 4.3 results in partitioning into four groups [A,B], [C,D], [E,F], [G,H] of the set of terminal elements.

Each of the clusters is completely described by the list of elements it contains. The choice of a cutting level in the clustering tree is an operation that should be guided by the values of the index: whenever possible, a cut should occur where there is a jump in the index.

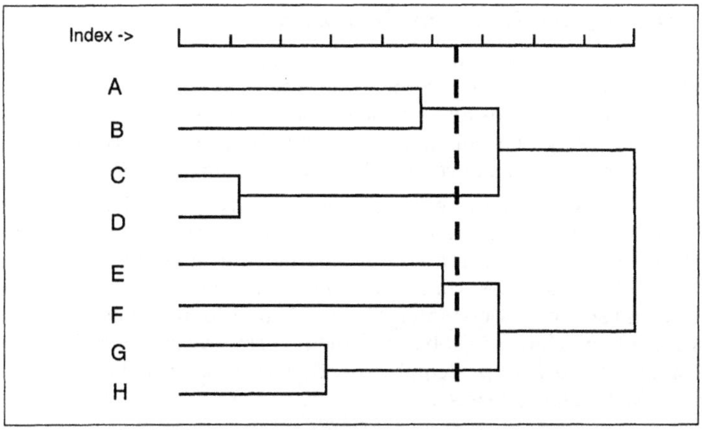

Figure 4.3
Cutting a dendrogram to obtain 4 groups

On figures 4.1 to 4.3, the values of the index to the left of the cut should be relatively small, and those to the right of the cut should be relatively large: this way elements within a cluster defined by the cut are similar to one another and those belonging to a distinct cluster are not similar to the first set. This topic will be revisited in section 4.3.1 which deals with methods for mixed clustering.

4.1.3 Appending supplementary elements

Starting with a hierarchical cluster analysis performed on a set of elements, it is possible to assign a set of supplementary or illustrative elements to clusters [1].

For each supplementary element, the appending algorithm proceeds in a particularly simple fashion. We start by searching among the elements of the basic set for the one that is closest to the supplementary element to be classified. Then this supplementary element is assigned to each cluster that contains the basic element.

This method can be used, as in the case of correspondence analysis, to illustrate the cluster set based on words with segment data. Here again the

1 This concept is being used with the strict meaning of the definition of supplementary elements introduced in the previous chapter.

method used allows us to classify a large number of supplementary elements without disrupting the clusters obtained on the basic set of elements (active elements).

4.1.4 Filtering on first principal axes

As mentioned above, hierarchical cluster analysis is built on a distance calculated among elements of the basic set considered pairwise. In the general case, the distances among the elements subjected to cluster analysis are measured using the chi-square distance between the columns of the table. This is the same distance used in correspondence analysis.

Recall that the goal of correspondence analysis is precisely to extract, whenever possible, subspaces that best summarize the information contained in the initial matrix. Interpretation usually relies on the hypothesis that the axes that correspond to the smallest eigenvalues constitute "noise" that is not interpretable. If one follows this logic, cluster analyses can be performed on basic elements using distances measured only in the space of the first principal axes, that is, those considered the most significant in a statistical sense.

The operation that transforms the distances measured among the basic elements and reduces them to their projection into the space of the first factors constitutes a *filtering* of the distances on the first axes. It makes it possible, through a selective reduction of information, to cluster thousands of words or individuals.[1]

This filtering property that is common to most methods based on finding principal axes (singular value decomposition, principal components analysis, simple and multiple correspondence analysis) is also used in *textual discriminant analysis* (chapter 8) and in information retrieval in the approach called *latent semantic indexing* (Deerwester et al., 1990; Bartell et al., 1992).

4.2 CLASSIFICATION OF ROWS AND COLUMNS OF A LEXICAL TABLE

As in the preceding chapter, the essentials of the method are presented by means of a small example, remembering that this pedagogical exercise is intended to demonstrate the method and thus cannot give an exact idea of the heuristic value of the tool. We shall look in succession at clustering rows (words) and columns (text parts).

1 A preliminary correspondence analysis has two additional advantages. First, it provides a complementary description of the data using different computational principles. Second, it can lead to significant time savings in the computations of distances in the case of large matrices (in information retrieval, for example).

4.2.1 Cluster analysis of words

We shall use the chapter 3 example that cross-tabulates *14* words and *5* survey respondent categories. The cluster analysis is performed on the rows of table 4.1 (which reproduces table 3.2 of the preceding chapter). The rows of this table are the row-profiles among which are calculated the distances of interest.

The first column of this table contains an order number that serves as an identifier for the clusters. The original bases (i.e., total counts) for each words are in the last column. Table 4.2 describes the steps and parameters of the algorithm.

Table 4.1
Row-profiles of table 3.2 (chapter 3)
(*Base* designates the count corresponding to each word)

		No Degree (1)	Elem. Sch. (2)	Trade Sch. (3)	High Sch. (4)	Coll. (5)	Total	(Base)
1	Money	26.4	33.2	16.6	15.0	8.8	100.	(193)
2	Future	16.7	28.3	24.5	23.6	6.9	100.	(318)
3	Unemployment	25.1	39.2	17.7	14.1	3.9	100.	(283)
4	Decision	4.5	31.8	22.7	22.7	18.2	100.	(22)
5	Difficult	25.9	40.7	14.8	11.1	7.4	100.	(27)
6	Economic	13.0	24.1	22.2	20.4	20.4	100.	(54)
7	Selfishness	19.6	34.6	13.1	24.3	8.4	100.	(107)
8	Occupation	15.2	44.3	24.1	7.6	8.9	100.	(79)
9	Finances	35.7	25.0	25.0	10.7	3.6	100.	(28)
10	War	15.4	26.9	26.9	23.1	7.7	100.	(26)
11	Housing	15.4	42.3	13.5	19.2	9.6	100.	(52)
12	Fear	15.7	28.3	23.9	23.9	8.2	100.	(159)
13	Health	19.4	29.0	21.5	20.4	9.7	100.	(93)
14	Work	23.2	40.4	19.2	9.3	7.9	100.	(151)
	Total	20.3	33.7	20.2	17.9	7.9	100.	(1592)

Reading table 4.2

- *First step:* There are *14* elements to cluster.
  The first row of table 4.2 indicates that the first artificial element (*node*) is obtained by joining elements 2 and 10 (first and last), the words *future* and *war*. The new element is number 15. It is characterized by the mean profile of its two components. The value of the corresponding index (*0.00008*) describes the smallest corresponding distance, and the weight of the new element (*344*) is the sum of the bases (*344 = 318 + 26*).

- *Second step:* 13 elements to cluster (second row of table 4.2).
  The two closest elements, that become element 16, are elements (*future + war*) and 12 (*fear*).
- *Third step*: 12 elements to cluster (third row of table 4.2)
  Element 17 is then formed by the union of elements 14 and 5 (*work, difficult*), element 18 joins the pair (*decision, economic*). Then (*selfishness, housing*), ([*future+war+fear*], and *health*), etc.
- The process ends when there remains only one element.

Table 4.2

Hierarchical cluster analysis of words (description of nodes)

NUM.	FIRST	LAST	BASE	WEIGHT	INDEX	HISTOGRAM OF LEVEL INDICES
15	2	10	2	344.	.00008	*
16	15	12	3	503.	.00018	*
17	14	5	2	178.	.00022	*
18	6	4	2	76.	.00061	**
19	7	11	2	159.	.00094	***
20	16	13	4	596.	.00107	***
21	8	17	3	257.	.00198	******
22	9	1	2	221.	.00219	******
23	3	22	3	504.	.00321	*********
24	21	23	6	761.	.00523	**************
25	20	19	6	755.	.00654	****************
26	25	18	8	831.	.00890	************************
27	26	24	14	1592.	.03091	*************************/ /***
SUM OF LEVEL INDICES =					.06206	

Note that the sum of the indices (*0.062*) is also equal to the sum of the eigenvalues calculated from the same data table in chapter 3. As seen previously, this quantity is proportional to the chi-square calculated on the contingency table.

Thus correspondence analysis and cluster analysis decompose the same quantity in two different ways (classic chi-square), and this quantity measures a deviation between the observed situation and the hypothesis of independence of the rows and columns of the table.

The graphical representation of this dendrogram illustrates this process of agglomeration more powerfully. We shall compare it to the graphical representation we obtained previously through correspondence analysis, so as to emphasize the originality of each point of view.

Figure 4.4 below displays the successive groupings of table 4.2 in the form of a dendrogram: the length of the branches of the tree is proportional to the values of the index.

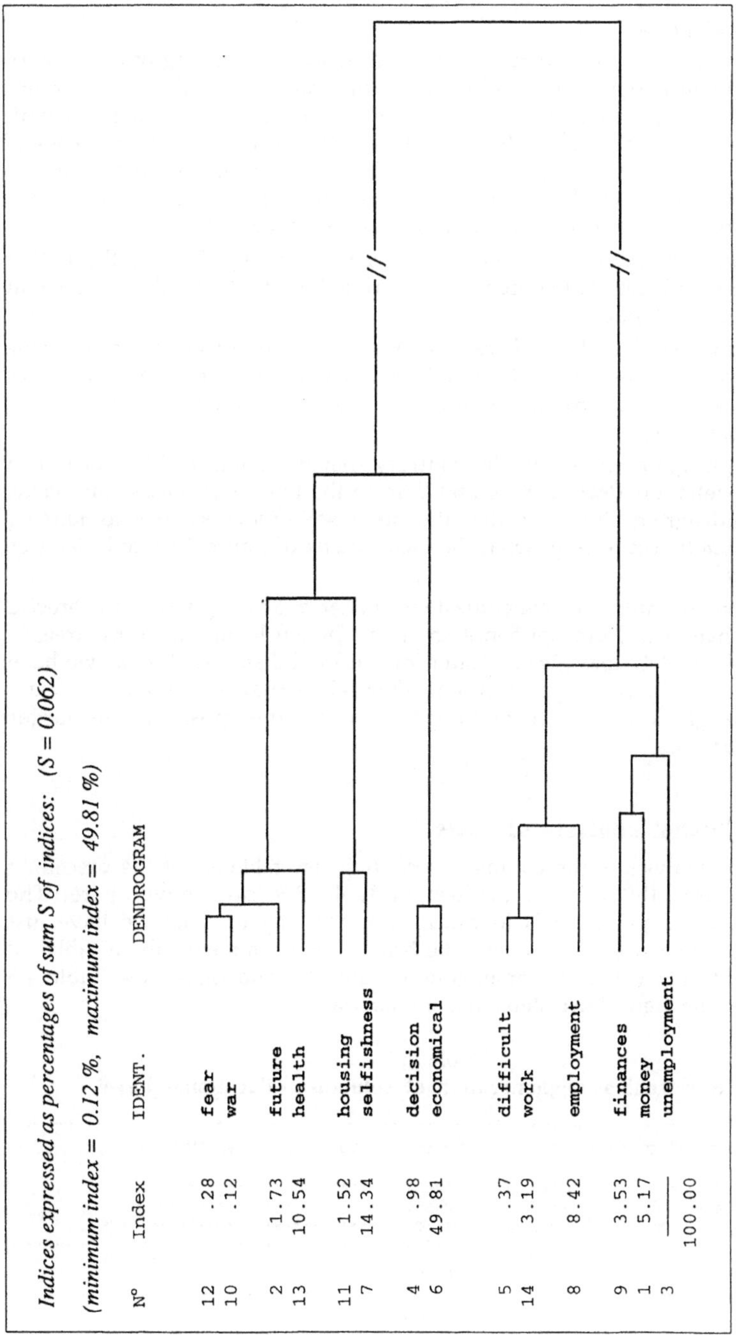

Figure 4.4: Dendrogram describing proximities among rows (words) of table 4.1

Reading figure 4.4

On this figure we read that the words *fear, war, future* agglomerate very early, but also that the "bundle" (*fear, war, future, health*) only becomes joined with the pair (*housing, selfishness*) much later. A comparison with figure 3.1 of the preceding chapter is interesting. The two main branches of the dendrogram, that oppose the eight first rows (from *fear* to *economical*) to the last six rows (from *difficult* to *unemployment*) translate the main polarization that is seen on the first (horizontal) principal axis.

The main groupings observed in the map of figure 3.1 are those that describe the process of hierarchical clustering, but with some differences that deserve our attention.

The points *health* and *selfishness* are close on figure 3.1 although it is true that they are close to the origin. Recall that in correspondence analysis proximities have a stronger significance when the points occupy positions on the periphery.

The dendrogram shows on the contrary that the point *health* is nearer to the constellation (*fear, war, future*) than to the point *selfishness*. Similarly, this dendrogram shows us that the point *selfishness* is closer to *housing* than to *health*, contrary to what the factor plane of figure 3.1 would lead us to believe.

However one must not conclude that cluster analysis gives more precise results than does correspondence analysis. The left hand part of the tree, in figures 4.1 to 4.4, does give a clearer idea of local distances; but, as we have seen, the branches can pivot around themselves (figure 4.2), and, in itself, the tree gives little information on the relative positions of larger constellations.

4.2.2 Cluster analysis of texts

A cluster analysis of the columns of contingency table 3.1 of the preceding chapter uses the column profiles of table 3.3, not shown here. The agglomeration principle is identical in every way to what we have just discussed for rows, and we shall be brief in our comments about table 4.3 and figure 4.5, which are analogous to table 4.2 and figure 4.4. Table 4.3 describes the algorithm's steps in a similar way.

Table 4.3
Hierarchical agglomeration of columns: Educational Level

NUM.	FIRST	LAST	BASE.	WEIGHT	INDEX	HISTOGRAM OF LEVEL INDICES
6	2	1	2	*860.00*	*.00816*	********************
7	4	3	2	*607.00*	*.00950*	************************
8	7	5	3	*732.00*	*.01171*	*****************************
9	8	6	5	*1592.00*	*.03269*	*****************************/ /*****
SUM OF LEVEL INDICES =					*.06206*	

There are 5 elements to cluster, and the first node, number 6, is formed from categories 2 and category 1 [columns number 2 (*Elem. School*) and 1 (*No Degree*) on table 4.1] and totalling *860* occurrences. Then the categories *High School* and *Trade School* are joined to category *College*.

Figure 4.5 outlines the process, showing as before that the two main branches of the tree correspond to an opposition along the first principal axis of figure 3.1 of the previous chapter.

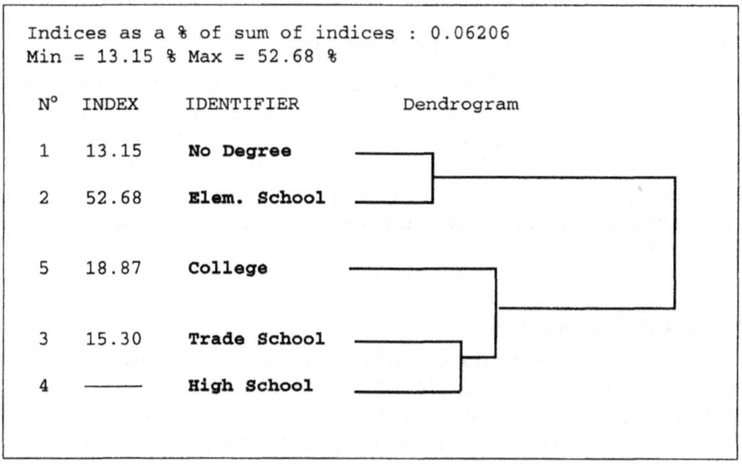

Figure 4.5
Dendrogram describing proximities among columns of table 3.1

Note that the sum of the indices is still 0.062, which is what we expect, because this quantity makes use of the table's rows and columns in a symmetric fashion. However, contrary to what happens in correspondence analysis, the cluster analysis of the columns cannot be simply deduced from the cluster analysis of the rows. There is no equivalent to the transition equations.

4.2.3 Notes on cluster analysis of words

Even though cluster analyses performed on a set of text parts (or respondents) and those done on a set of words (or segments) appear to be linked from a formal viewpoint, the two types of analyses meet different needs, leading to different utilizations of the method.

When texts are being analyzed (literary, political, historical texts), the number of words is generally far greater than the number of text parts. The dendrogram has so many terminal elements that any global summarization becomes rather complicated. In practice, the interpretation of such a large cluster analysis is performed by considering in order of priority the associations that appear at the two extremities of the dendrogram:

- Clusters at the lower levels of the hierarchy comprising agglomerations of words with a very small index (i.e. clusters that become agglomerated at the outset of the clustering process)
- Higher level clusters, often comprising many words, that are analyzed as entities. At higher levels of clusters, the main observed oppositions of the first principal plane (spanned by the first two principal axes) are generally found.

Associations at the lower levels of the hierarchy

First, we are concerned with associations that occur at the first levels of the cluster analysis. By construction, these associations group together sets of words whose distribution profiles are very similar in the parts of the corpus. As we shall see, it is necessary to examine which associations are mostly repetitions of segments, which ones are repeated co-occurrences of several words within the same sentences or paragraphs, and which ones might occur by chance.

An example of "quasi-segments"[1]

In a corpus of labor union texts, consisting of seven texts originating from four labor union shops[2] (U1, U2, U3, U4), the distribution of the words *weighing* and *heavily* showed perfectly identical distributions of the two words, resulting in their agglomeration right at the beginning of the cluster analysis.

Table 4.4
Distribution of two words in four texts

	U1	U2	U3	U4
heavily	0	3	7	1
weighing	0	3	7	1

This identity suggests the presence of occurrences of the stereotype "*weighing heavily (on something or someone)*" in different parts of the corpus. A search in the context was carried out to check for the existence of this phrasing in several parts of the corpus. It was observed that the co-occurrences of these two words did not occur exclusively in this one context.

An inspection of the distributions for these two words shows first of all that within this corpus they are used in contexts of the following type:

1 Cf. also Becue (1993) whose terminology we are borrowing.
2 This corpus of main French labor union shops texts (between 1971 and 1976) has been analysed by Bergougnioux *et al.* (1982) and Salem (1993).

> [U2] /.../ Local governments have taken over maintaining green spaces and this is *weighing heavily* on beneficiaries of public housing programs.
>
> [U3] /.../ a new direction in economic policy /.../ is necessary to deal with inflation which is *weighing heavily* on workers and low income families as well as having negative effects on foreign trade.

In addition to these contexts, the two words also appear in slightly different contexts.

> [U2] /.../ the national committee /.../ has observed that retail prices are continuing to rise and that inflation is *weighing* more and more *heavily* on all low income social categories, mostly the elderly and heads of households.

Finally each of these two words appears in contexts that are independent from occurrences of the other word: sometimes we find other words that are associated with a particular word that compensate for the absence of the word *heavily*. This is the case for this context of the word *weighing*:

> [U2] /.../ so that Europe can be *weighing* in *as favorably as possible* in opposition to the two major economic blocks

or:

> [U3] /.../ the strength of the American economy is *weighing a great deal* on other economies.

The identical distributions of the two words call for two comments. First, the overall resemblance of the profiles results essentially from the existence of a recurring and more or less rigid expression that contains both words. Second, the exactly identical distributions are somewhat due to chance in this example, since they are the result of adding occurrences associated to this expression to other occurrences of the two words coming from separate contexts. Clustering methods make it possible to generalize the search for co-occurrences of pairs of words within a sentence to finding co-occurrences that can concern several words in larger contexts.

Clusters containing homogeneous distributions

Words are clustered into a small number of groups by cutting the tree at a high enough level in the indexed hierarchy of the clusters. This process provides an opportunity to consider clusters that group together words whose distribution is relatively homogeneous. The content of these clusters is analyzed by thoroughly examining the context in an attempt to find meaningful reasons for the observed similarities among the profiles of words. This procedure also provides a way of verifying that occurrences of text units (words or graphical forms or types) that correspond to different inflections of the same lemma are assigned to the same cluster, reflecting an analogous distribution in different parts of the corpus.

Documentation by segmental data

Here again, the readability of the clusters obtained on words increases significantly if distributions of repeated segments are added to the results shown on words.

As in the case of correspondence analysis, segments can be included in two different ways: first, by direct analysis of the table of words and repeated segments on which all of the elements are included as active elements; second, by including the segments as supplementary (illustrative) elements after a cluster analysis performed on words. The algorithm for adding supplementary elements to a cluster analysis (cf. Jambu, 1978) adds a set of statistical individuals that were not part of the cluster analysis to the various clusters of the indexed hierarchy, without disturbing the cluster analysis. We shall make use of this property to describe cluster analyses with illustrative repeated segments.

Hierarchical cluster analysis can also be performed on the combination of elements (words and repeated segments) whose frequency is greater than a certain threshold in the corpus. Although this second manner of proceeding does give different results from those obtained using the words alone, experience has shown that the two procedures usually lead to a similar interpretation. The new clusters then contain words and segments whose distribution is relatively homogeneous in the parts of the corpus. The analysis of the relative position of a segment and the words it contains is expedited through this procedure.

Influence of partitioning the corpus

We have seen that the distances among the different words are calculated on the basis of their distribution in the parts of the corpus. It follows that the manner in which the corpus is divided into parts becomes an important issue in the construction of a set of clusters based on the words within one corpus. The reason is that variations in how a corpus is partitioned can bring certain pairs of words closer and pull others apart, thus affecting the hierarchical cluster analysis. Thanks to the growing capacity of computers in terms of memory and computational speed, hierarchical cluster methods can be used to find textual co-occurrences within sentences, as well as fixed length text segments and paragraphs. The algorithms described here should be able to stretch research capabilities by working on ever finer partitions of a corpus.

4.3 CLUSTER ANALYSIS OF SURVEY DATA SETS

In analyzing survey data sets, the theoretical and practical difficulties of a process that uses principal axes methods are intensified: they have to deal with both the nature of the results and their sheer volume. We shall emphasize two issues:

- Visualizations are often limited to a small number of dimensions (most often two) whereas the actual dimensionality of the phenomenon under study can be much higher (this dimensionality is measured by the number of significant axes[1]).

- These visualizations can include several hundred points, and give rise to crowded or illegible graphs, and to lengthy lists of coordinates.

Thus it is important to make use of the data management and computational capabilities of the computer to complete and clarify the presentation of the results. The combined use of clustering methods and correspondence analysis can fill in the gaps. When there are too many points on a graph it may be useful to group the data into homogeneous families. The algorithms used for developing these groups work the same way whether the points are located in a two-dimensional or a ten-dimensional space. In other words, the process has two objectives: to minimize graphical printouts, on the one hand, and to work with the real dimensionality of the configuration of points, on the other hand.

Once the individuals have been grouped into clusters, it is straightforward to obtain a description of these clusters: indeed, statistics related to differences between internal values for each cluster and overall values for the sample can be calculated for numerical variables and categorical variables. These statistics can also be converted into *test-values* and sorted on these test-values. Finally, the most characteristic response categories and variables can be displayed for each cluster.

After a brief review of clustering algorithms that are suitable for large data sets, we shall present an application example that is an extension of multiple correspondence analysis presented in the preceding chapter (section 3.3). This example illustrates this new way of presenting results in a compact fashion.

4.3.1 Mixed clustering algorithms

The clustering algorithm that seems most appropriate for partitioning a data set comprising thousands of individuals is a *mixed algorithm* consisting of four phases:

a)*Initial partitioning* into several dozen groups through a "dynamic configurations" or "k-means" type technique (MacQueen, 1967; Diday, 1971). These techniques can be summarized as follows. First a few individuals are drawn at random. They become provisional cluster centroids. Then all of the individuals are assigned to the nearest provisional centroid (using a distance such as the chi-square distance defined in the preceding chapter). Thus a partition of the set of individuals is constructed. Provisional centers are calculated again, and they are now the "centroids" (mean points, for example) of the clusters just developed. This process is reiterated, in other words all of the

[1] From a statistical viewpoint, a *significant axis* is an axis that explains an amount of variance that cannot be imputed to chance.

individuals are assigned again to these centroids, resulting in a new partition, etc. This process eventually stabilizes, but the partition obtained generally depends on the initial choice of centroids.[1]

b) *Hierarchical agglomeration of clusters*: The hierarchical clustering techniques presented in this chapter are rather costly when applied to several thousand elements. For this reason it is important to reduce the dimensionality of the data by performing a preliminary grouping into several dozen clusters. The dendrogram is important because it can give an idea of the actual number of clusters in the population. Each cutting of this tree yields a partition consisting of ever fewer clusters as one cuts closer to the root of the dendrogram.

c) *Cutting the tree* (generally after visual inspection). The more points are agglomerated, in other words, the closer one gets to the top of the tree, the greater is the distance between the two closest clusters. We saw (section 4.1.2) that by cutting the tree at the level of a significant jump in the index, the chances of obtaining a good quality partition are greater, because the individuals grouped together before the jump are similar, and those grouped together after the cut are much less similar, which is a practical definition of a good partition.

d) *Optimization of the partition by re-assignments.* A partition that is obtained by cutting is not the best possible one, because the algorithm for hierarchical clustering unfortunately does not have the property of giving an optimized partition at each step. The partition can be improved again by re-assigning the individuals as indicated in paragraph *a* above. Despite the relative complexity of the procedure, one can still not be assured of having found the "best partitioning into *k* clusters".

4.3.2 Sequence of operations in survey analysis

Full analyses that include a filtering phase through correspondence analysis proceed in the following manner:

a) Selecting active elements (somewhat equivalent to choosing a point of view). Survey respondents can be described from the viewpoint of their basic characteristics, or in terms of a particular theme: product usage, political opinions, magazine readership, etc.

b) Simple or multiple correspondence analysis using these active elements.

c) Positioning of illustrative elements. All available information is projected in order to understand or interpret the typology obtained from the active elements.

---

1 In finding an optimal partition with respect to a criterion, such as those commonly used in statistics (for example, maximizing the *between variance/within variance* quotient), we are still confronted with prohibitively high computational costs. This problem has yet to be solved, even with the most powerful computers.

d) Inspection of the planar maps, generally limited to points that occupy the most significant positions.

e) Partitioning of the set of individuals or observations according to the method described above.

f) Positioning on the preceding maps of the centroids of the main clusters (each partition becomes a nominal variable).

g) Description of clusters by the most characteristic response categories and variables.

4.3.3 Application example: working demographic partition

This example is an extension of the multiple correspondence analysis of the preceding chapter. The *144* survey respondents are characterized by a set of active variables. Now the goal is to obtain a small number of clusters of individuals that are as homogeneous as possible with respect to their basic characteristics. The significance of such groups will become clear in chapter 5, where open-ended responses are agglomerated without the use of any particular criteria.

Ideally, one would like to be able to cross-tabulate characteristics such as age, gender, profession and educational level, and thus create groups of individuals that can be exactly compared to one another on the basis of their basic characteristics, that is, to produce a *ceteris paribus* analysis. In practice such cross-tabulations very quickly generate thousands of categories, an unwieldy number when dealing with a sample of approximately *1000* individuals. In addition, the cross-tabulations do not take into account the network of interrelationships that may exist among these characteristics. Some of these associations are straightforward (there are no young retired persons), others are known but might have exceptions (there are few widowed students or college educated workers). Finally some have a more statistical basis (there are more women in the employed and widowed categories).

As a rule one can expect to find operational groupings of approximately *20* clusters for a sample on the order of *2000* individuals. We shall see that a more general grouping into five large clusters is somewhat viable, while remaining compatible with the restricted format of a pedagogical example. Such clusters constitute a *working demographic partition (WDP)*. They are used in chapter 5 to create groupings of open-ended responses (section 5.2).

Application to example of section 3.3.1

The mixed classification procedure described above was applied to the *144* individuals using the five active variables described in section 3.3.1. Four clusters were obtained by cutting the dendrogram, followed by optimization. Here we shall present only a new step: description of the clusters.

Guide to reading table 4.5

The first two columns of numbers (V-TEST and PROB), are the most important ones, since they make it possible to select and rank the most characteristic categories of each cluster.

For each cluster, the most characteristic categories are listed in order of decreasing values of the test-value V-TEST, or, equivalently, to increasing values of the probability PROB (second column).

The test-value V-TEST is, briefly, the analog of the value of a standardized normal variable, which is significant (at the *0.05* level) if it lies between the values *-1.96* and *+1.96.* Because of the problem of *multiple comparisons*, the quantity V-TEST is considered here as a pragmatic criterion, interesting only if its value is quite large (about multiple comparisons, see also chapter 6, section 6.1.2). All of the categories shown here are significant.

The GRP/CAT numerical column shows the percentages of each cluster in the categories (row-percents): thus all of the widow(er)s in the sample belong to cluster 4 (GRP/CAT = *100*).

The CAT/GRP numerical column gives the internal percentage in each of the cluster solution's categories (column percents). Thus there are *20%* widow(er)s in cluster 4 (CAT/GRP=*20*). In this case the internal percentage is higher than the overall percentage (column : OVERALL) since the selected categories characterize the clusters. The difference between the internal percentage and the overall percentage is the basis for computing the test-value.

Thus cluster 3 contains *38* individuals, or *26%* of the population. The category "Management", which applies to *5.56%* of the sample, pertains to *15.79%* of the persons of this cluster. With a test-value of *2.62* it is ranked fourth in describing this cluster.

The OVERALL column displays the percentage of each category in the overall population. It is obtained by dividing the column WEIGHT by the sample base of *144*.

The WEIGHT column shows the raw counts of the clusters and categories.

Table 4.5

Example of description of a partition into 4 *working demographic groups* (example of section 3.3.1)

CHARACTERISTIC CATEGORIES		V.TEST	PROB	----- PERCENTAGES ---- GRP/CAT	CAT/GRP	OVERALL	WEIGHT
- GROUP 1 / 4						45.83	66
Educational Level	High School	7.61	.000	93.02	60.61	29.86	43
Educational Level	Trade School	3.50	.000	72.22	39.39	25.00	36
- GROUP 2 / 4						13.89	20
Educational Level	None or Elem.	8.82	.000	74.07	100.00	18.75	27
Profession	Employee	2.58	.005	42.86	30.00	9.72	14
Age in 3 cat.	30 - 50 years	2.33	.010	24.49	60.00	34.03	49
- GROUP 3 / 4						26.39	38
Educational Level	College	12.42	.000	100.00	100.00	26.39	38
Age en 3 classes	less than 30	2.66	.004	36.49	71.05	51.39	74
Profession	Management Prof.	2.62	.004	75.00	15.79	5.56	8
Profession	Student	2.42	.008	40.43	50.00	32.64	47
- GROUP 4 / 4						13.89	20
Age in 3 cat.	Greater than 50	9.94	.000	95.24	100.00	14.58	21
Children	Yes	3.59	.000	28.85	75.00	36.11	52
Marital Status	Widowed	3.45	.000	100.00	20.00	2.78	4
Gender	Female	2.57	.005	21.62	80.00	51.39	74
Profession	Housewife	2.44	.007	57.14	20.00	4.86	7
Educational Level	Trade School	2.39	.008	27.78	50.00	25.00	36
Profession	Craftsman, Storekeeper	2.39	.009	75.00	15.00	2.78	4

Table 4.5 describes the clusters in an exact way, by comparing the percentages of responses within the clusters to the overall sample percentages, then by selecting the most characteristic categories (cf. guide for reading table 4.5). Note that among the most characteristic categories of a cluster along with the active categories, are found the illustrative categories (those that did not participate in the formation of the clusters). Such is the case of categories that describe professions.

The advantages of creating partitions to describe multidimensional sets are rather obvious:

The concept of a cluster is intuitive (groups of individuals that are as similar as possible). The description of the clusters makes use of fully labelled, easy to read descriptors. These descriptions are based on simple comparisons of percentages. Thus it is easier to describe clusters than a continuous space.

But correspondence analysis achieves a visualization of the relative positions of clusters in space, and reveals continuous or anomalous variations in this space that could be masked by the discontinuity of the clusters. The two techniques complement one another and are mutually validating.

5

VISUALIZATION OF TEXTUAL DATA

How do we go about applying the multivariate techniques defined through the pedagogical examples in the preceding chapters to real-life situations? The information is so complex, and the possible points of view so numerous, that it is impossible to recommend a single path leading from the problem to a definitive solution. In this chapter we shall rather attempt to recognize different ways of delaying somewhat the moment when the user must necessarily intercede in an interpretative manner. Our aim, briefly, is to extend the scope of the analysis that is controllable and reproducible - we choose to use these simple words instead of the perhaps more controversial terms *objective* and *automatic*.

Our purpose here is to help to analyse a series of texts, whether they be literary texts, documents, responses to open-ended questions grouped into artificial texts (groupings based on age categories, profession, educational level or any other relevant criterion).

Which texts are most similar with respect to vocabulary and frequency of use of words (in other words, which texts have similar lexical profiles)? Which words are characteristic of each text, either through their presence or absence? The reader may recognize these questions as the types of questions that may be answered with correspondence analysis of the *lexical table* (the table that cross-tabulates words and texts).

This type of analysis is developed in section 5.1 under the title: *Correspondence analysis of lexical tables.*

In the case of open-ended responses in surveys, the approach we propose assumes that the responses have already been grouped. But the most pertinent grouping criteria are not known a priori. Also, it is not always possible to try all feasible combinations, due to the number of variables that could be used for grouping, and the even greater number of cross-tabulations of these variables.

We propose two analytic strategies when there is no specific categorical variable present to create text groupings:

a) Using a generated partition. An automatic clustering technique already outlined in chapter 4 can be used to summarize the various characteristics of respondents into a single partition. An example of such a *Working Demographic Partition* is developed in section 5.2.

b) Direct analysis of ungrouped responses. If the responses appear to be rich enough, a direct correspondence analysis of the lexical table that cross-tabulates words and responses can be performed. Such an analysis produces graphical displays of the responses (which are generally quite rough) and dual displays of the words. Responses belong to respondents, and, in most cases, the characteristics of these respondents are known thanks to their other responses to closed questions asked during the same survey. Thus it is possible to project these characteristics onto the displays as supplementary elements.

The problems and elements of solutions relative to this direct approach are discussed in section 5.3, entitled: *Direct analysis of responses or documents.*

These various analyses are presented through examples[1]. Most of them refer to the same corpus of responses to an open-ended question.

5.1 CORRESPONDENCE ANALYSIS OF LEXICAL TABLES

We saw in the preceding chapters that responses can be coded numerically in a way that is completely "transparent" to the user. The result of this numerical coding can take two different formats, coded in two tables **R** and T.

5.1.1 Basic lexical tables

Table **R** has as many rows as there are responses. In general this number of rows (number of individuals) is noted k, (there can be missing responses, but it is convenient to reserve a row for each respondent, to ensure uncomplicated merging with other information).

Table **R** also has a number of columns equal to the length of the longest response (i.e. the number of occurrences in this response). For individual i, row i of table **R** contains the addresses of the words that constitute his/her response, while respecting the order and the possible repetitions of these

[1] Another possible approach is the *Correspondence analysis of juxtaposed lexical tables.* In this case, a lexical table is built for each nominal variable and the juxtaposition of these tables is analyzed. This method is a straightforward generalization of the method developed in section 5.1.

words. These addresses refer to the vocabulary that is inherent to the response. Table **R** thus makes it possible to reconstitute the original responses integrally. In practice table **R** is not rectangular, because each row is of variable length. The integers of which table **R** is composed cannot be bigger than V, which is the size of the vocabulary.

Table **T** has the same number of rows as table **R** but it has as many columns (V) as there are words (or segments) used by all of the individuals. The cell defined by row i and column j of **T** contains the number of times word j is used by individual i in his/her response. Thus this is a contingency table (responses × words), that is, a lexical table. Table **T** can easily be constructed on the basis of table **R**, but the converse is not true: information relative to the order of the words in each response is lost in table **T**. Actually, table **R** is much more compact than table **T**[1].

5.1.2 Aggregated lexical tables

In most applications isolated responses are too sparse to be the object of a direct statistical treatment: it is necessary to work on grouped responses[2].

Using the notation of chapter 3, we designate by **Z** the complete disjunctive table with k rows and p columns that describes the responses of k individuals to a closed-end question with p possible response categories, where the responses are mutually exclusive. In other words, each row of **Z** has only one "*1*", and *(p-1)* "*0*"s.

Unlike a table such as table 3.9, table **Z** includes only one question, and thus is a single block. Each closed-end question of this type defines a partition of the respondents.

Table **C**, obtained through the matrix product:

$$\mathbf{C} = \mathbf{T'Z}$$

is a table with V rows (recall that V is the total number of distinct words) and p columns (p is the number of response categories for the closed-end question under consideration) whose general term c_{ij} is none other than the number of times word i is used by the set of individuals having chosen response j.

An example of table **C** of small size is given in table 3.1 of chapter 3, with $V=14$, and $p=5$. In this case the closed-end question was: *What is the highest educational degree you have achieved?*

---

1 Thus, a response containing *20* occurrences (for a lexicon of *1000* words) generates a row of length *20* in table **R** and a row of length *1000* in table **T** (this row has at least *980* zeroes). In fact the statistical and algorithmic treatments that involve table **T** are programmed with the help of table **R**, which uses less of the computer's memory.

2 See section 5.3.2 for the direct analysis of a sparse matrix **T**.

Thus it is straightforward to calculate the aggregated lexical table $\mathbf{C_q}$ for any closed-end question whose responses are coded into a table $\mathbf{Z_q}$ with the equation:

$$\mathbf{C_q} = \mathbf{T'Z_q}$$

The same holds true for comparing the lexical profiles of different categories of the population. Each table $\mathbf{C_q}$ offers a different viewpoint (i.e. the viewpoint of the closed-end question q on the distribution of the lexical profiles of the responses to the open-ended question being analyzed).

5.1.3 Frequency threshold for words

These comparisons of lexical profiles only become meaningful from a statistical point of view if the words appear with a certain minimum frequency: hapaxes or even rare words are eliminated from the phase of comparisons of frequencies.

As a first consequence, the size of the vocabulary taken into account is considerably reduced. Frequency distributions of vocabularies are such that choosing a frequency threshold often drastically reduces the size of the vocabulary [1].

As a second consequence, patterns obtained are more significant and statistical tests are more powerful. It is a classical result that the accuracy of most estimators (estimators of the mean or the median of usual distributions, estimators of a proportion) increases with the size of the sample [2].

5.1.4 Example

The example below serves as an illustration of the major steps involved in the statistical processing. The following open-ended question was asked:

"What is the single most important thing in life for you?" It was followed by the probe: *"What other things are very important to you?"*.

This question was included in a multinational survey conducted in five countries (Japan, France, Germany, United Kingdom, USA) in the late nineteen eighties (Hayashi *et al.*, 1992).

---

1 For example, for one open-ended question and *2000* respondents, if only words that appear at least *10* times are selected (i.e. a frequency threshold of *10*), the value of V is often reduced from *1000* to *100*.

2 In most cases, the accuracy is proportional to the square root of the sample size. In the case of patterns obtained in the space spanned by the two first dimensions in correspondence analysis of a contingency table, a simple (rough) test can be performed on the significance of the distance of each point to the origin. For a row-point i whose frequency is $k_{i.}$, such a distance must be greater than $r_i = \sqrt{5.99 / k_{i.}}$, (*5.99* is the value given by the chi-square table for $p=0.05$ and for 2 degrees of freedom) (see: Lebart et al., 1984). Thus, the accuracy of the location of an element increases with its frequency.

Table 5.1
Words Appearing at Least Sixteen Times (Alphabetic Order) in the *1043* responses to the open question

Word	Frequency	Word	Frequency	Word	Frequency
I	*248*	go	*19*	of	*312*
I'm	22	going	*26*	on	*59*
a	*298*	good	*303*	other	*33*
able	*55*	grandchildren	*30*	others	*17*
about	*31*	happiness	227	our	*29*
after	*26*	happy	137	out	*34*
all	*86*	have	*99*	own	*16*
and	*504*	having	*70*	peace	*77*
anything	*19*	health	*609*	people	*61*
are	*65*	healthy	*45*	really	*28*
as	*83*	help	*20*	satisfaction	*21*
at	*29*	holidays	*17*	security	*40*
be	*136*	home	*90*	see	*26*
being	*116*	house	*34*	should	*23*
can	*35*	husband	*86*	so	*21*
can't	*19*	if	*18*	son	*25*
car	*18*	important	*26*	standard	*33*
children	*123*	in	*180*	suppose	*23*
church	*16*	is	*57*	that	*97*
comfortable	*20*	it	*61*	that's	*62*
comfortably	*17*	job	*142*	the	*331*
content	*18*	just	*52*	their	*21*
contentment	*31*	keep	*50*	them	*25*
daughter	*19*	keeping	*37*	they	*24*
day	*16*	kids	*16*	things	*28*
do	*47*	leisure	*16*	think	*45*
dog	*16*	life	*161*	time	*29*
don't	*35*	like	*40*	to	*521*
education	*25*	live	*84*	up	*29*
else	*47*	living	*68*	very	*19*
employment	*20*	long	*16*	want	*31*
enjoy	*17*	love	*33*	way	*17*
enough	*68*	make	*16*	we	*38*
everything	*16*	me	*33*	welfare	*22*
family	*684*	mind	*47*	well	*59*
family's	*16*	money	*170*	what	*22*
food	*23*	more	*21*	which	*17*
for	*107*	much	*17*	wife	*68*
freedom	*38*	music	*18*	with	*89*
friends	*116*	my	*810*	work	*116*
from	*17*	myself	*21*	world	*29*
future	*17*	nice	*30*	worries	*16*
general	*25*	no	*46*	would	*19*
get	*54*	not	*51*	you	*36*
getting	*35*	nothing	*56*	your	*21*

Our illustrative example is limited to the British sample (Sample size: *1043*). This multinational survey concerning *general social attitudes* is described in detail in Sasaki and Suzuki (1989).

The same questionnaire also had a number of closed-end questions (among them, the socio-demographic characteristics of the respondents, which play a major role in the discussions that follow). In this example we focus on a partitioning of the sample into nine categories, obtained by cross-tabulating age (three categories) with educational level (three categories).

Several examples of actual responses to the same open-ended question are shown in chapter 2 (table 2.1) and in chapter 6 (tables 6.5 and 6.6, which are displays of a selection of characteristic responses).

Numeric coding phase

The counts for the first phase of numeric coding are as follows:

Out of *1043* responses, there are *13 669* occurrences, with *1 413* distinct words. When the words appearing at least *16* times are selected, there remain *10 357* occurrences of these words, with *135* distinct words (types).

Table 5.1 shows the list in alphabetical order of the *135* words that appear at least *16* times in the sample of *1043* responses to the question mentioned above. Note that graphical forms such as *can't, don't, that's,* are considered as words because the apostrophe is not designated as a separator in this example. This alphabetical listing of the words shows that some words are connected to a single dictionary word *(be, is, are, being)*. Also to be noted is the strong presence of function words *(a, and, for, that, the)*. As we mentioned previously, consolidating and cleaning out the vocabulary is not strictly necessary if the objective of the analysis is only to detect differences among groups.

If the *function words* are distributed in random fashion among the various groups of individuals, they do not have much effect on the results of exploratory statistical analyses (such as correspondence analysis). Conversely, if they are not random (as frequently observed), they may be an interesting part of the analysis. Similarly, if two inflections of the same lemma behave identically, they can just as well be replaced by a single word (property of *distributional equivalence* of the chi-square distance used in correspondence analysis). If they behave differently, then they refer to different contexts of use of the word, and this difference deserves to be noted.

Table 5.2 shows the various segments observed for the responses in our example, sorted in alphabetical order of the first word that appears in the segment, and selected only through frequency thresholds: segments of length two (two consecutive words) are very numerous, whereas they often have little to offer from a semantic point of view. Segments of length two must appear at least *50* times to be included in the inventory, and segments of length three or more must appear at least six times.

Table 5.2
Partial list of segments (global threshold: *6*, threshold for *segments of length 2: 50*, threshold for *segments of length 3: 6*)

Number	Frequency	Length	Text of Segment
			(I)
1	*10*	*3*	I like to
			(a)
2	*54*	*2*	a good
3	*8*	*3*	a good life
4	*7*	*3*	a nice home
			(able)
5	*52*	*2*	able to
6	*7*	*3*	able to get
7	*9*	*3*	able to live
			(as)
8	*11*	*3*	as long as
			(be)
9	*26*	*3*	be able to
10	*9*	*3*	be happy to
			(being)
11	*23*	*3*	being able to
			(can't)
12	*10*	*4*	can't think of anything
13	*8*	*5*	can't think of anything else
			(enough)
14	*34*	*3*	enough money to
15	*16*	*4*	enough money to live
16	*9*	*5*	enough money to live on
			(family)
17	*60*	*2*	family health
18	*73*	*2*	family my
19	*15*	*3*	family and friends
20	*11*	*3*	family good health
21	*7*	*3*	family health happiness
22	*13*	*3*	family my health
23	*10*	*3*	family my job
24	*9*	*3*	family my work
			(good)
25	*176*	*2*	good health
26	*8*	*3*	good health and
27	*17*	*3*	good health family
28	*9*	*3*	good health for
29	*14*	*3*	good health happiness
30	*15*	*3*	good health my
31	*10*	*4*	good standard of living
			(happy)
32	*13*	*3*	happy family life
			(have)
33	*10*	*3*	have a good
			(having)
34	*13*	*3*	having enough money
35	*8*	*4*	having enough money to

5.1.5 Construction of aggregated lexical and segmental table

The numeric coding of free responses is completely transparent to the user, since the input for most software packages is actual responses in text form. As seen previously, the result of this coding can be summarized in table **T**. Table **T** has k rows (number of respondents, $k=1043$ in our example) and as many columns as there are words (or segments) used by the entire set of individuals, that is, V columns ($V=135$, or $V=81$ if we work with segments). At the intersection of row i and column j of table **T** is the number of times $t(i, j)$ word j is used by individual i in his/her response. Thus this is an 'individual × words' contingency table.

In most applications, single responses are too sparse to be used for statistical processing. It is necessary to work with responses that have been grouped together on the basis of nominal variables measured on the same individuals. Aggregating the k rows of **T** into m respondent categories yields a contingency table **C**.

Table **C** serves to compare the lexical (or segmental) profiles of various segments of the population. These comparisons of profiles make sense from a statistical point of view only when the words appear with a great enough frequency; words that only appear once (hapaxes) or rarely occurring words are discarded. As mentioned above, a suitable frequency threshold has the effect of reducing the size V of the vocabulary. As a particular consequence of the highly asymmetric shape of the frequency distribution of words (see section 2.3.2 of chapter 2), *13* percent of the distinct words account for *76* percent of the text. Responses to the open-ended question of our example are generally very short. Because of this, two answers that seem at first to refer to similar contents, may have no words in common, and therefore may not be recognized as similar.

This is an important issue which we shall address several times in this chapter and in the following chapters. Briefly, everything depends on the degree of refinement desired in the analysis, as well as on the sample size, which are closely linked. A quick content analysis on a small sample encounters problems of distributions of words relative to the same lemma, and also synonymy problems. The same is not true of large response sets. In the latter case, larger frequencies make it possible to validate the usages of terms or phrases, and sometimes to identify categories of respondents that use them preferentially.

Thus it is appropriate to find as an initial step groupings of individuals that are pertinent to the phenomenon being analyzed. We will discuss later how such a grouping is found. In the particular case of our example, the individuals are grouped into 9 categories that differ with respect to age or educational level.

This partitioning into nine groups is obtained by cross-tabulating two partitions, each of which comprises three categories. First partition: three age categories (less than 30 years [noted -30], between 30 years and 55 years [-55] over 55 years [+. 55]). Second partition: Three educational levels (No degree or Low [noted L], Medium [M], and High level [H]).

Table 5.3

Partial listing of lexical table cross-tabulating 135 words of frequency greater than or equal to 16 with 9 age-education categories

	L-30	L-55	L+55	M-30	M-55	M+55	H-30	H-55	H+55
I	2	46	92	30	25	19	11	21	2
I'm	2	5	9	3	2	1	0	0	0
a	10	56	66	54	44	19	20	22	7
able	1	9	16	9	7	4	4	5	0
about	0	3	13	7	1	2	4	1	0
after	1	8	11	3	1	2	0	0	0
all	1	24	19	8	18	6	3	5	2
and	8	89	148	86	73	30	25	32	13
anything	0	4	9	1	3	0	1	1	0
are	0	18	24	4	13	2	1	3	0
as	1	15	38	9	10	4	4	1	1
at	0	8	7	4	5	2	3	0	0
be	3	32	40	21	20	4	9	4	3
being	2	17	25	27	17	5	13	9	1
can	2	5	17	4	2	3	2	0	0
can't	0	6	8	1	2	0	1	1	0
car	1	0	6	6	2	0	1	2	0
children	2	36	24	18	30	3	3	6	1
church	0	3	3	0	1	4	1	4	0
comfortable	0	4	7	3	3	0	2	1	0
comfortably	0	3	6	5	2	0	0	1	0
content	0	5	4	2	4	1	1	0	1
contentment	0	7	5	3	8	2	3	2	1
daughter	0	4	5	4	1	3	1	0	1
day	0	1	9	1	2	0	1	0	2
do	0	8	15	9	4	0	8	3	0
dog	0	4	5	3	2	1	0	1	0
don't	1	7	11	6	3	2	2	2	1
education	2	4	1	4	6	0	4	3	1
else	0	8	13	9	6	3	6	1	1
employment	2	5	3	3	4	2	1	0	0
enjoy	1	4	4	1	4	1	0	2	0
enough	1	14	16	10	17	4	3	2	1
everything	0	5	3	2	3	1	2	0	0
family	11	177	139	115	118	40	30	39	15
family's	0	4	5	3	2	1	1	0	0
food	0	2	7	2	4	3	4	1	0
for	0	27	30	8	20	9	2	8	3
freedom	0	3	4	9	12	3	3	4	0
friends	3	7	28	23	18	9	17	9	2

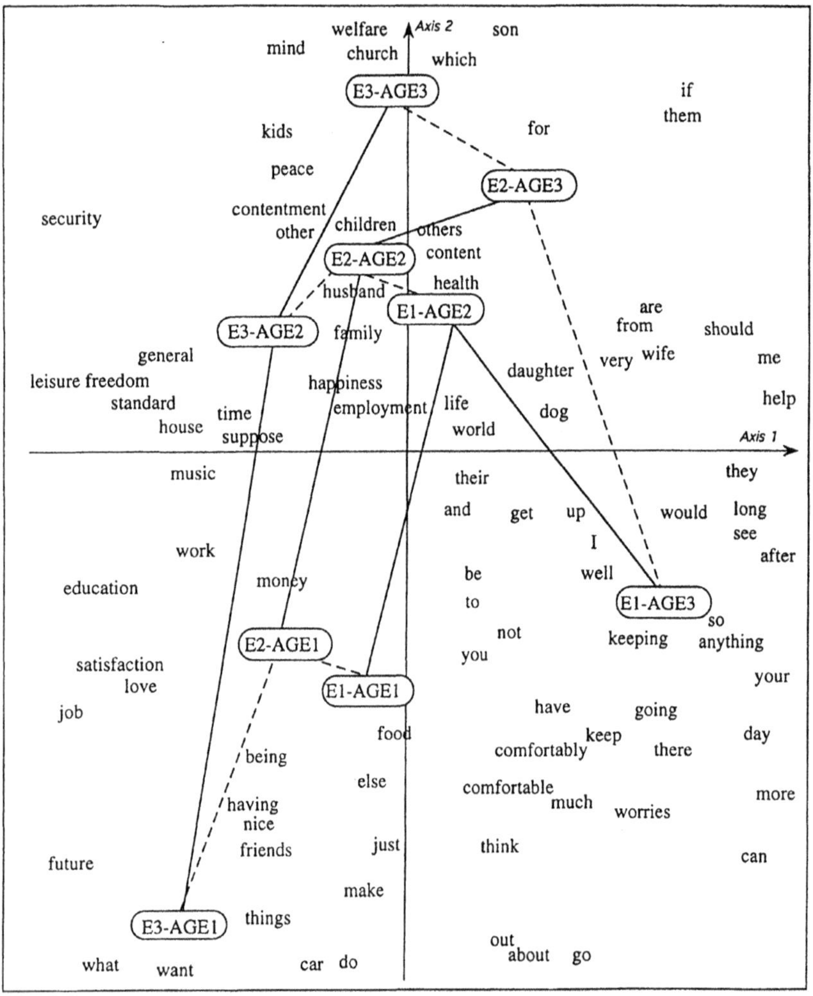

Legend			
E1:	Education Low	AGE1:	<30
E2:	Education Medium	AGE2:	30-55
E3:	Education High	AGE3:	>55

Figure 5.1
Correspondence analysis of table 5.3 (first principal plane) (only a subset of words has been plotted)

To analyse the information contained in this table effectively, the row-profile and column-profile tables are calculated, and the distances among words on the one hand, and age-education categories on the other, are displayed. It is precisely the purpose of correspondence analysis to provide the user with such a dual description.

5.1.6 Analysis and interpretation of lexical tables

Figure 5.1 shows the first principal plane[1] (that is, the plane of the first two principal axes) of the correspondence analysis of table 5.3. Only a subset of the 135 active words has been plotted. The selection criterion for this subset is the *quality of representation* in the plane[2].

The first two eigenvalues are respectively *0.046* and *0.029*, and account for *28.3* % and *17.9* % of the trace (or total variance).

The arrangement of the column-points is remarkably regular: on the basis of purely lexical information (elements of column profiles) the composite character of the partitioning of the individuals into *9* categories is recreated. At the same age, individuals with a higher educational level are situated toward the left of the graph; at the same educational level, the older they are, the higher they appear along the vertical axis. Thus, these vectors that describe the frequency of *135* words (chosen according to a simple frequency criterion) for each category can reconstitute simultaneously the gradations of ages (within each educational level category), and the gradations of educational levels (within each age category).

Interpreting the positioning of the points that represent the words is, however, limited by the absence of context of these points at the present level of analysis.

It can be seen on figure 5.1 that the highest educational levels are positioned on the left hand side. They use the words *security, leisure, freedom, peace, mind* also appearing on the left side of the graph.

The older categories are positioned on the upper right side of the graph. Except for the group 'E1-AGE3' (Low education, age over *55*), educational level increases from the right to the left, and age increases from the bottom to the top.

Also note that the two sets of "younger" age levels cluster more densely than the older age level points. This phenomenon is often observed in a general socioeconomic setting: there tends to be a greater range of life situations (living standard, educational level) among older people than among younger people.

---

1 On figure 5.1, categories belonging to the same level of education are connected by a bold line, and categories belonging to the same age category are connected by a dashed line.

2 The quality of representation of an element i is the square of the following ratio: *distance of i to the origin in the plane* divided by *distance of i to the origin in the whole space*. It is nothing but the sum of the squared correlations of i corresponding to the two first axes (see section 3.2.3 and appendix 1). This criterion attains its maximum value of *1* if the element belongs to the plane.

Figure 5.2 is an example of the role played by external information. Words belonging to a specific part of speech, nouns, are in bold.

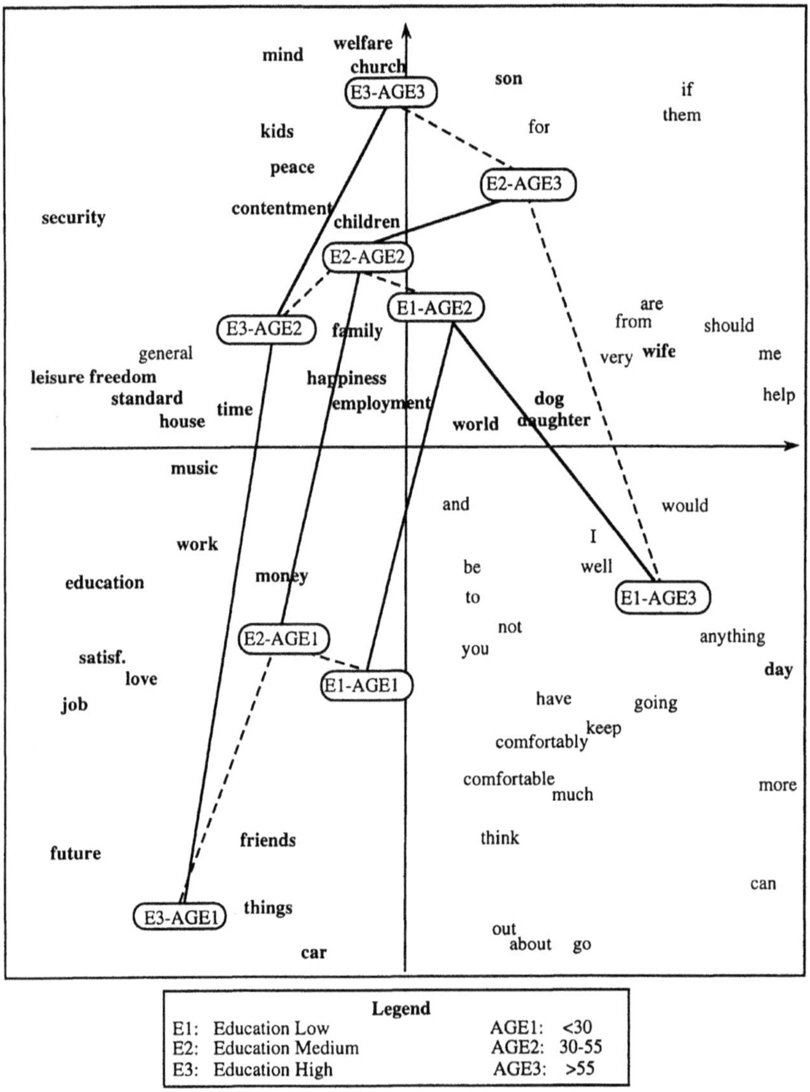

Figure 5.2

Correspondence analysis of table 5.3 : Nouns are in bold letters (same selection of words as figure 5.1)

Clearly, nouns are not scattered at random in this plane: they are concentrated on the left side, and characteristic of educated categories (see, for instance, Somers, 1966).

This socio-linguistic result can be enhanced through further modulations of the original display: adjectives, verbs and pronouns could be identified. It can be seen that verbs and personal pronouns are more frequent in the responses of older persons with a low educational level.

The graphical display can also be enriched by identifying the words according to general semantic categories. For example, it appears that all words related to the concept of family (*wife, children, kids, son, daughter*) characterize older respondents, whatever their educational level.

Interpretation

Several comments are relevant at this point:

a) The automatic indexing of words and frequency computations ignore on purpose much information of a semantic or syntactic nature that is available to any reader. Neither synonyms nor homonyms are accounted for. Applications of this type of analysis to large samples (*1043* responses in our example) show that these objections can easily be waived in the case of artificial texts constructed by juxtaposition of responses, where the main purpose is to find repeated elements.

   In this statistical context, analyzing words often gives more interesting results than analyzing lemmas or groups of words established on a semantic basis. The words *comfortable* and *comfortably* occupy similar positions in figure 5.1 (lower right) which suggests that keeping these two words distinct is not a problem. The frequency threshold chosen (forms appearing at least *16* times) eliminates the word *relationship* that appears *15* times, as well as its plural form appearing *12* times. An analysis with a lower threshold, or positioning these words as a supplementary element shows that these two words would almost be superimposed on figure 5.1. In such a case, the property of *distributional equivalence* of the chi-square distance used in correspondence analysis ensures a stability of the results, whether or not these words are merged.

   In other application examples, we have noted that the presence of different forms of one verb (such as: *can, able to, could*) and of synonyms helps confirm the interpretation of certain zones of the principal plane: a grouping of different words that have semantic affinities is a validation criterion for empirical results.

b) Including function words (such as *for, if, then, from,...)* does not burden the analysis in any way. These words are seldom uniformly distributed among different kinds of respondents. Consequently, their profiles could frequently depart from the average profile (origin of the axes), placing them in significant positions on a correspondence analysis map.

c) The order of the words in the responses is not taken into account: each utterance becomes a "bag of words" for computer programs, and only frequency profiles are actually analyzed. This is a serious objection, although frequency profiles have shown in practice to be much richer sources of information than one would imagine *a priori*. Whereas a lexical profile, that is, a series of *135* sub-frequencies does not have much significance on its own, a comparison of several lexical profiles is on the contrary abundant in information.

Again with analyses based upon frequencies, finding *repeated segments* makes it possible to take into account occurrences of units that are richer at the semantic level than isolated words. The selection of *modal responses* which is discussed later, also highlights the more frequent contexts of some of these words.

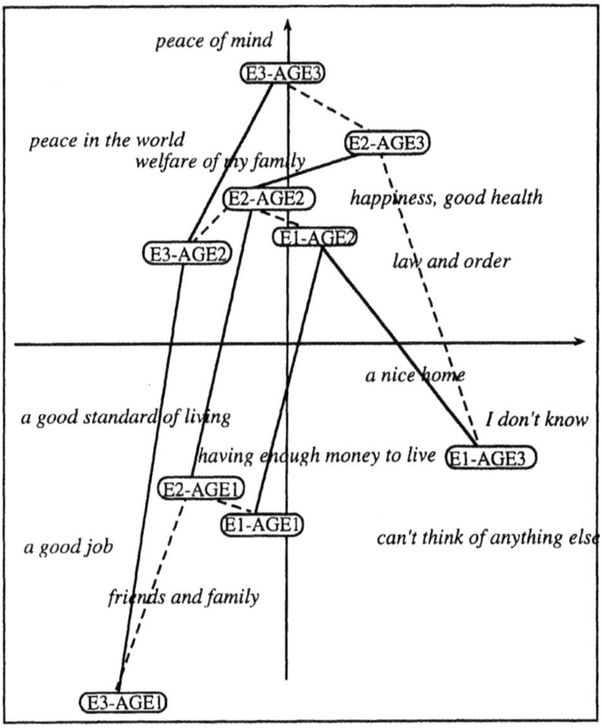

Figure 5.3

Positioning some segments as illustrative elements (same principal plane as figures 5.1 and 5.2)

5.1.7 Illustration of displays using repeated segments

As shown in chapter 3, maps such as figure 5.1 can be enriched with supplementary (or illustrative) elements. The lexical and the segmental profiles can both be represented as points in the same p-dimensional space. The lexical table **C**, with V rows (forms or words) and p columns (categories), and a segmental table $\mathbf{C}_S$, with s rows (segments) and p columns (categories) are stacked together. We can then project the segmental profiles onto the principal plane obtained previously with the words as *active elements*. This generally leads to rather cumbersome graphical displays that are much more manageable interactively on a computer screen than in book format.

Figure 5.3 shows a selection of meaningful segments overlaid on the map of figure 5.1. Some words now appear within their contexts of repeated segments, and the real semantic content of the responses begins to emerge from this purely blind and automated processing.

5.1.8 Stability vis-à-vis an internal lemmatization

Remarks a) and b) of the preceding paragraph raise the question of the stability of the results when a list of function words is eliminated and/or the words are systematically replaced by the lemmas to which they are linked.

An internal lemmatization procedure applied on the sub-corpus obtained after establishing a threshold makes it possible to evaluate the stability of the structures obtained.

The question is to verify whether the observed pattern (relative positioning of nine category points on figure 5.1) is an artifact due to the presence of particular grammatical words. If that were the case, the categories would be distinguished primarily through their use of certain parts of speech, and not solely through the content of their responses.

Grouping several words increases the frequency of the resulting aggregated element. Thus, it is necessary to lower the frequency threshold of words.[1] We begin with a list of *296* words corresponding to the threshold *6* (forms appearing at least *6* times). In this list, we have eliminated the following words:

---

[1] For instance, the lemma *relationship* corresponds to two graphical forms whose frequencies were less than 16 (*relationship* and *relationships*). These graphical forms were eliminated from the list of table 5.1. The frequency of the resulting lemma is now greater than the initial threshold. Hence the importance of lemmatizing before applying this threshold.

Eliminated words:

a, an, and, at, but, by, etc., for, from, if, in, into, of, on, or, out, over, than, the, to, up.

Additionally we have combined into single units the following word sets (in each row the first word remains and replaces the one or several others)

Words (or forms) declared as equivalent to the first one in each row:

be	←	am, m, are, re, is, been, being, was,
child	←	children,
content	←	contented,
can	←	could,
would	←	d,
do	←	doing,
enjoy	←	enjoying,
family	←	families,
get	←	got, getting,
go	←	going,
have	←	having, ve,
help	←	helping,
holiday	←	holidays,
job	←	jobs,
keep	←	keeping,
live	←	living,
look	←	looking,
relationship	←	relationships
see	←	seeing,
son	←	sons,
sport	←	sports,
thing	←	things,
work	←	working,
worry	←	worries,

This is a sort of internal lemmatization of the selected set of words, (after specifying a sub-corpus by choosing a threshold (6) based on frequencies of words) and not a real lemmatization like the one presented in chapter 2 (section 2.7). It is clear that when the lemmatization is done on the entire original text, it also has an influence on the constitution of the set of words selected via a threshold.

In particular, verbs, which have many more inflections than nouns and adjectives in the romance languages, are at a disadvantage in a selection based on frequencies of words, since many usages of one verb can appear in the original text with a frequency less than the threshold.

Nevertheless, the point of view selected for this experiment is to calculate the internal stability of the selected sub-corpus. Here the vocabulary is reduced from 296 words to 240 "pseudo-lemmas". In this particular example, the map of figure 5.1 is not strongly affected by this considerable reduction in the number of words.

If we keep the *143* words appearing at least *12* times in this transformed corpus, we obtain a new lexical table analogous to that of table 5.3.

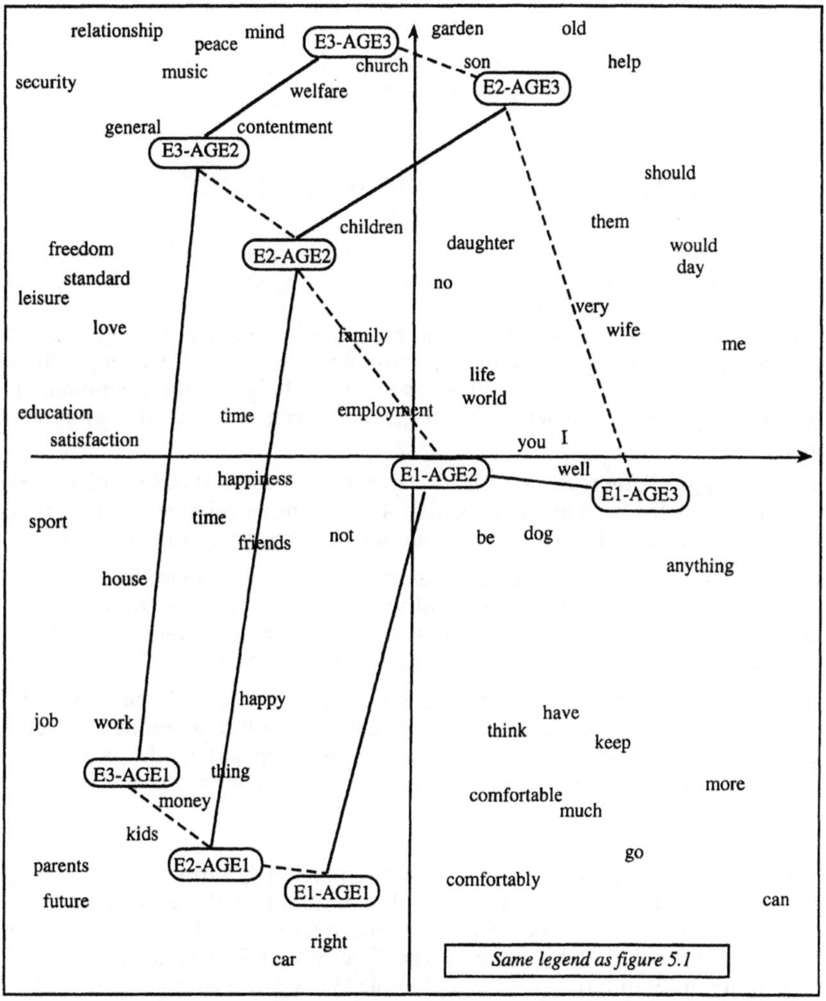

Figure 5.4

Proximities between words and categories of respondents after an internal lemmatization of the corpus of responses

A description of this new table through correspondence analysis produces a pattern of category-points similar to that of figure 5.1 (the patterns of both words and categories are sketched on figure 5.4). Such a transformation of the data set allows the user to assess the stability of the results obtained.

We have seen that the category-points are positioned in a way that respects the order of the age and educational level categories. This adds to the conjecture (but does not prove) that there is a connection between the categories and the content of the responses.

5.2 WORKING DEMOGRAPHIC PARTITIONS

As we have seen, it is often necessary to group responses before proceeding to statistical analyses. The lexical profiles of aggregated responses are smoother and are more readily interpretable than isolated responses. The preceding examples have shown how such an *a priori* grouping can be implemented with the variables chosen on the basis of various initial hypotheses. However this implies a solid knowledge of the phenomenon under study, a situation which is not generally true in so-called exploratory analyses.

The technique known as *working demographic partitions (WDP)* which was discussed briefly in chapter 4 (section 4.3.3 example relative to clustering techniques), is useful in providing material for addressing this problem.

Given a list of descriptors or variables that characterize the individuals, the problem is to allocate these individuals into groups that are as homogeneous as possible with respect to the chosen characteristics, without placing excessive emphasis on any one characteristic a priori.

As we stated in chapter 4, the goal is to approach a *ceteris paribus* state within each group, a situation that is typically difficult to attain in the social sciences. This is what a clustering algorithm is meant to do as well as possible, as discussed in the example given at the end of chapter 4.

An example

The example that follows concerns an open-ended question included in a survey about the living conditions and aspirations of the French (in 1980 and 1981). *4 000* verbatim responses are available. The questionnaire touched on the following themes: The Family and Family Relations, The Environment, Standards of Living, Life at Work and some general topics: Science, Justice, Social Life, Leisure.

The wording of the open-ended question was as follows:

> *You have just answered many questions concerning your life in general as well as your environment. Perhaps you would like to give your opinion about some topics that were not included in this questionnaire. Do you have anything of your own you would like to add?*

Our objective is to have an overall view of the responses. However the variety of subjects covered in this survey, and thus the assortment of the possible contents of the responses makes it impossible to know beforehand which are the most pertinent criteria for grouping the responses.

Starting with a battery of *13* descriptors listed below, the *4 000* surveyed individuals are grouped into *22* groups (including a "residual" group, representing less than *3%* of the sample, that contains incomplete questionnaires, or a heterogeneous set of low frequency combinations of characteristics).

List of descriptors *(active variables in cluster analysis)*

	Number of categories
Socio-professional category	15
Marital status	6
Highest educational degree	8
Present employment	8
Presence of children <16 at home	2
Occupational status of household	5
Gender-work status cross-tab	4
Size of town	5
Number of persons in household	5
Gender-age Cross-tab	8
Household income	7
Ownership of dishwasher	2
Ownership of several cars	2

Figure 5.5 shows the cutting of the hierarchical tree for *22* final groups and gives a summary description of the composition of each of these groups.

Figure 5.6 sketches the principal plane obtained by performing a correspondence analysis on the lexical table that cross-tabulates the 150 highest frequency words as rows and the *22* groups as columns. Only a few words are shown on this figure.

This partition highlights peculiarities and oppositions that might have been ignored if simpler groupings had been used.

Thus better educated persons give critical evaluations of the questionnaire, whereas older persons tend to mention problems that concern them particularly (*retirement, taxes, etc.).*

Nevertheless, some nuances that are more difficult to see may appear. Note, for example, a considerable difference in lexical profiles among male workers who are childless (group 17, lower left part of the display) and those with children (group 18, upper right part).

The latter, obviously interested and motivated, raise issues that are not covered in the questionnaire, or add to discussions initiated during the interview, whereas the former are contented to make rather laconic comments on the questionnaire.

Percentage	*Description*	*Number*
(7.8%)	Unemp., M, 2 in HH, elderly	1
(1.7%)	Unemp., M, 1 in HH, elderly	2
(5.2%)	Unemp., F, 1 in HH, elderly,	3
(3.3%)	Unemp., F, 2 in HH, elderly,	4
(4.2%)	Unemp., M&F, 1 in HH, (student)	5
(2.6%)	Unemp., M&F, ill or invalid	6
(5.5%)	Unemp., F, mid. age, no children	7
(2.3%)	Unemp., no child., former H.wives	8
(6.0%)	Unemp., child., young, low income	9
(3.0%)	Unemp., child., young, high income	10
(1.8%)	Empl., F, children,young, high inc.	11
(2.7%)	Empl., M, children, young	12
(2.8%)	Empl., M, no children, young	13
(5.6%)	Empl., M, children, mid. age	14
(3.6%)	Empl., F, child., mid. age, farmer	15
(4.3%)	Empl., M, child., mid. age, farmer	16
(7.4%)	Empl., no chil., mid. age, laborer	17
(10.4%)	Empl., children, mid. age, laborer	18
(5.1%)	Empl., M&F, single, young	19
(4.1%)	Empl., F, no child., mid.age	20
(7.6%)	Empl., F, children, young	21
(2.8%)	Residual group	22

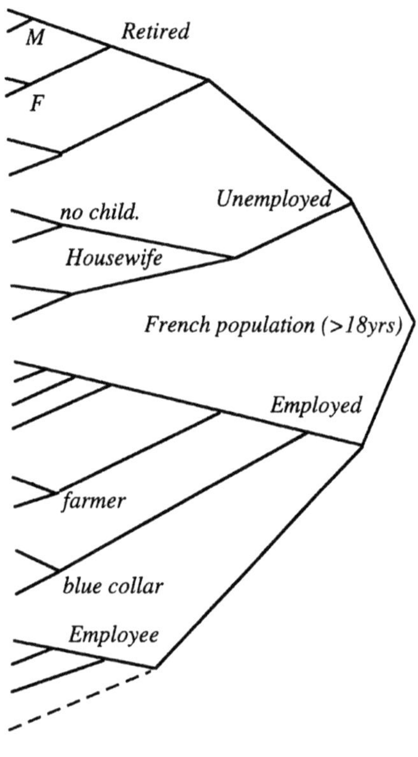

Legend

F:	*Female,*	M:	*Male,*
HH:	*Household,*	H.wives :	*Housewives,*
Empl. :	*Employed,*	Unemp. :	*Unemployed.*

Figure 5.5

Partitioning into *22* Working Demographic Clusters (from a sample of *4 000* persons)

The grouping of the 22 groups seen on figure 5.6 give an idea of the underlying variables. Nodes 1, 2, 3, 4 and 8 at the bottom right are all made of elderly persons, whereas nodes 12 and 13 are formed of employed young males. Groups 9, 10 and 21, made up of females, often mention that certain problems of family life are missing from the questionnaire.

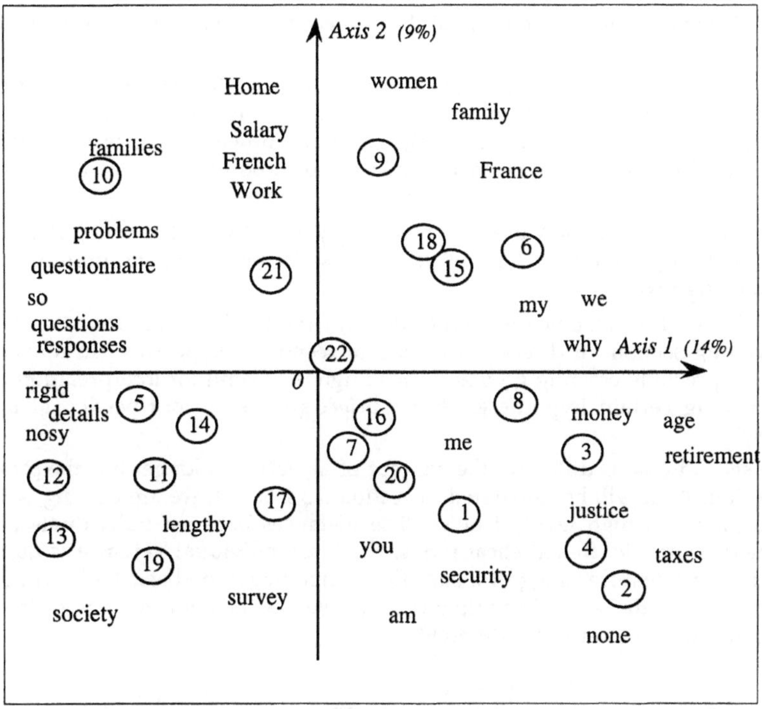

Figure 5.6

Proximities among *22* working clusters and a selection of words (Question: *comments at end of interview*)

Concerning this same open-ended question, the techniques of modal responses presented in chapter 6 make it possible to place the words shown on this figure in their contexts, and thus to considerably enrich the interpretation of these groups.

5.3 DIRECT ANALYSIS OF RESPONSES OR DOCUMENTS

A major portion of our discussion has been concerned with analyzing frequency profiles for relatively lengthy texts. To obtain such texts from open-ended response files, we have proceeded thus far to *a priori* groupings of these responses. However some of the exploratory statistical methods presented in chapters 3 and 4 can be applied effectively to individual responses.

This direct treatment of individual data is recommended in two instances as follows:

a) When the responses are sufficiently rich lexically that the frequency profiles can be compared, which may be the case in lengthy interviews, in psychology or medicine for example, or in the processing of textual units that are fairly voluminous, in the area of socio-political text analysis, for example.

b) When the responses are not necessarily lengthy, but the objective is limited to a preliminary descriptive analysis aiming at creating groups and organizing text.

Clearly the first instance is applicable with the methods advocated in the preceding sections: a direct description of responses is possible. In no way does it preclude creating groups if they might be useful for interpretation or for testing certain hypotheses. But *a priori* groups are not needed at the outset.

The second case is different: the concept of a profile no longer has the same meaning. As it will be illustrated in section 5.3.1 below, we are dealing with data having a high level of noise. The distances between individuals are somewhat erratic. In statistical terms, the inter-individual variance is quite large. These features suggest a specific methodology: the basic tools will be the same, but the way of handling them as well as the interpretation of their outcome will be noticeably different.

5.3.1 How are distances interpreted?

We begin by taking a simple example of responses to an open-ended question about highway safety that follows a preliminary closed-end question:

Do you believe that it is possible to significantly lower the number of people killed and wounded in highway accidents? (responses *yes/no),*

then to those who answered *yes* (about *80%* in a 1982 survey):

What should be done to accomplish this?

The following types of responses are found:

- *increase use of public transportation*

or: - *motivate people to take trains and buses as much as possible*

These two responses do not have any words in common, and yet their contents are rather similar.

Conversely, the two following responses to the same question:

- *respect speed limits*
- *make people respect speed limits*

only differ by two words, and yet have significantly different contents.

This is a real life example, but a rather exceptional one. When aggregated lexical tables are treated, responses of this type are generally buried in groups whose mean lexical profile is quite regular. Be that as it may, these examples show that distances among individual responses cannot be interpreted easily.

Our objective, however, is not to exhaustively code the information, but to use redundancies and repetitions, when they exist, to simplify this information.

Clearly, direct analysis can at least group together identical or similar responses, while leaving aside for the time being responses that are distinctive because of the originality of their form. This can be considered as an aid to post-coding, the manual operation discussed in chapter 1, section 1.4.2.

5.3.2 Analysis of sparse matrix T

This direct analysis of responses consists of performing a descriptive analysis of table **T** defined above (section 5.1) whose k rows are responses and V columns are a set of words[1].

Several comments are in order.

a) The proximity between two words, that is, between two columns of matrix **T**, is greater when these words appear more often in the same responses (and not only in the same texts or the same response aggregates), which improves the ability to uncover syntagmatic proximities. Direct analysis gives a better accounting of contexts.

b) The transition equations of chapter 3 (section 3.2) provide a rather simple interpretation of the maps resulting from the correspondence analysis of matrix **T**: the k individuals are approximately at the centers of gravity of the words they use and conversely these words are positioned approximately at the centers of gravity of the individuals that chose them to express themselves (we use the term approximately to account for the presence of a dilation coefficient β for each axis, which moves the position of the centers of gravity).

c) The k rows of table **T** represent the responses or the individuals. The responses to the questionnaire's closed-end questions by the same k individuals can constitute the columns of a table $\mathbf{T}^+$, and thus be positioned as illustrative (or supplementary) elements on the principal axes planes resulting from the correspondence analysis of **T**. This makes possible a very rapid visualization of the proximities between the words and the

1 The treatment of such a large matrix generally involves using special computational algorithms, that use the reduced matrix **R** instead of matrix **T**, thus avoiding computing and storing a matrix to diagonalize of order (V, V) (cf. Lebart, 1982a).

characteristics of the respondents. In practice, it suggests possible grouping criteria. The practicalities of this type of exploratory approach are developed through an example.

5.3.3 Application example

In this example, matrix **T** cross-tabulates *1043* respondents (rows) and the *143* words (columns) appearing at least *12* times in the transformed corpus (after the internal lemmatization of section 5.1.8). Matrix $\mathbf{T}^+$ of supplementary elements is a complete disjunctive table containing *1043* rows and *29* columns (*6* blocks of *6* categorical variables : gender (*2* columns), educational level (*3* columns) age (*3* columns), gender × age (*6* columns) gender × education (*6* columns), education × age (*9* columns).

Why use a lemmatized version of the corpus?

Direct analysis of individual responses or documents generally performs better when applied to texts after a preliminary processing such as the one carried out in section 5.1.8. A grouping of words into lemmas reduces the number of units and consequently increases their frequencies. Therefore, the number of common units between two responses or documents also increases, and the corresponding distances computed between these responses become all the more meaningful.

The elimination of function words has a different effect : it has been shown previously that the role of these units is generally limited in the case of large texts (see the example of section 5.1.4). On the contrary, because of their high frequency, function words might play a major and unwelcome role when dealing with short responses or documents, inducing questionable proximities between responses.

At the risk of over-simplifying, we can say that function words introduce a "noise" into the individual responses at about the same level as the "signal" (the content). In agglomerating these responses into groups, the relative level of the noise is reduced (since the respondents speak the same language, use the same grammar, etc.). As mentioned before, if the profiles of function words differ from one group to another, it could be of interest to the researcher (from a socio-linguistic standpoint) since it means that there exist language differences among groups.

The map of words (figure 5.7)

The principal plane represented on figure 5.7 describes the proximities between words (columns of table **T**). Only a sample of the *143* words has been plotted. The rows of **T** (*1043* anonymous respondents) are not of interest to us. They are used to compute the positions of the supplementary categorical variables to be represented on a separate display (figure 5.8). Only responses containing at least three words have been kept as active elements for this analysis.

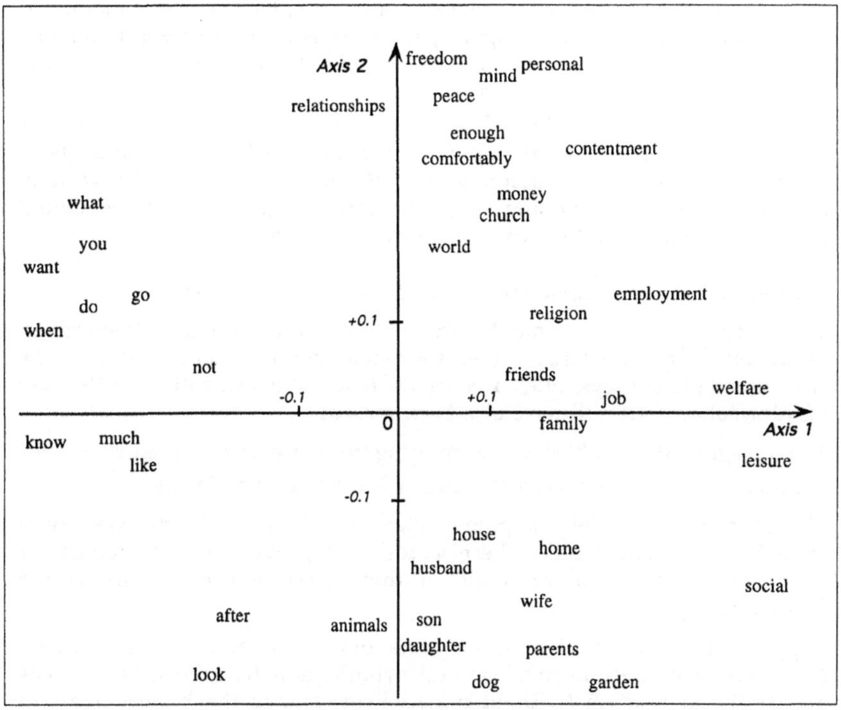

Figure 5.7
Positions of some words in the principal plane of a correspondence analysis of the sparse table T (words × responses) after internal lemmatization

As usual when dealing with large sparse matrices, the eigenvalues related to the first axes account for a minor part of the global variance (the eigenvalues are $\lambda_1 = 0.35$ and $\lambda_2 = 0.28$, the corresponding percentages being $\tau_1 = 2.4\%$ and $\tau_2 = 1.9\%$). These percentages, highly significant with respect to the hypothesis of independence between words and responses, must not be interpreted as percentages of information (see chapter 3)[1].

It has been stressed that these sorts of direct analyses are very coarse, due to the high level of noise in the individual responses or documents. It is clear,

1 It is useful to complement a direct analysis (and by doing so, to remedy the insufficiency of the visualizations obtained) with a clustering of words taking into account the distances in a much higher dimensional space.

however, that the groupings of words observed on figure 5.7 are far from being random. On the lower part of the display, we have a conglomeration of words relating to the *house* (and also *home, garden*) and its inhabitants. Beside the words *look, after,* one can find *wife, husband, son, daughter, parents, animals, dog.* On the upper part, there is a series of words relating to abstract concepts, values, and more general domains: *freedom, mind, personal, peace, relationships,* etc.

Obviously, these two clusters of words are seldom encountered in the same responses. They probably correspond to separate clusters of respondents. A set of verbs and adverbs occupies the left hand part of the map, without apparent content. Positioning supplementary categorical variables should help us understand and interpret this scatter diagram.

External validation: Supplementary variables and test values

The respondents (or documents) are not interesting in and of themselves. However, their characteristics (i.e., socio-demographic characteristics in the case of sample surveys; titles of journals, fields of investigation, in the case of information retrieval) have also been recorded.

The categorical variables corresponding to these characteristics in our example constitute the blocks of matrix $\mathbf{T}^+$ described previously.

The coordinates of these supplementary variables can be assessed with statistical parameters called "test-values" that provide a measurement of their significance. The principle of this assessment procedure is the following:

Suppose that a supplementary category j contains k_j individuals (respondents or documents). The null hypothesis is that the k_j individuals are chosen at random (without replacement) among the k individuals in the analysis.

Under these circumstances, on a specific principal axis α, the abscissa $\varphi_{\alpha j}$ of a supplementary category concerning k_j individuals is a random variable, with mean 0, and variance $v(j,k_j)$. Then, $t_\alpha(j) = \varphi_{\alpha j}/v(j,k_j)$ is a standardized random variable (having mean 0 and variance 1). Moreover, $t_\alpha(j)$ is asymptotically normally distributed[1]. Thus, using a common approximation, a test-value $t_\alpha(j)$ greater than 2 or less than -2 indicates a significant position of the corresponding category j on axis α (at level 0.05).

Table 5.4 presents the coordinates and test-values of the 29 supplementary categories. We can read for instance that the position of the variable gender is very significant on the first axis (test value = 6.0 for male), but not on the second one (test-value = 1.6).

1 The variance is the same whatever the axis under consideration, (see Lebart *et al.*, 1984). It is given by the formula: $v(j,k_j) = \dfrac{k-1}{k-k_j}\,\dfrac{1}{k_j}$.

Most of the supplementary categories have significant positions on these two first axes.

Table 5.4
Positioning categories of respondents on figure 5.7: Coordinates and test-values

	Coordinates		Test -values	
Gender				
male	*.07*	*.02*	*6.0*	*1.6*
female	*-.06*	*-.01*	*-6.0*	*-1.6*
Education				
low	*-.17*	*-.08*	*-15.3*	*-7.0*
medium	*.20*	*.00*	*14.1*	*.2*
high	*.07*	*.29*	*2.4*	*10.1*
Age 3 classes				
less than 30	*.10*	*.03*	*5.2*	*1.4*
from 30 to 55	*.13*	*.00*	*10.8*	*.1*
over 55	*-.24*	*-.02*	*-15.9*	*-1.4*
Gender*age				
male*-30	*.10*	*.07*	*3.2*	*2.4*
male*3055	*.32*	*-.03*	*12.7*	*-1.3*
male* +55	*-.18*	*.03*	*-7.3*	*1.4*
female*-30	*.11*	*-.01*	*3.7*	*-.5*
female*3055	*.03*	*.02*	*1.6*	*1.2*
female*+55	*-.30*	*-.07*	*-12.7*	*-3.0*
gender*education				
male*low	*-.09*	*-.09*	*-4.5*	*-4.3*
male*medium	*.24*	*.04*	*9.7*	*1.7*
male*high	*.21*	*.33*	*5.0*	*7.8*
female*low	*-.22*	*-.07*	*-12.8*	*-3.9*
female*medium	*.16*	*-.03*	*7.9*	*-1.3*
female*high	*-.07*	*.25*	*-1.6*	*6.0*
educ*age				
-30 * low	*-.14*	*.14*	*-1.6*	*1.7*
30-55 * low	*.09*	*-.12*	*4.3*	*-5.6*
+55 * low	*-.38*	*-.06*	*-20.9*	*-3.2*
-30 * medium	*.21*	*-.08*	*8.3*	*-3.1*
30-55 * medium	*.19*	*.09*	*7.5*	*3.5*
+55 * medium	*.19*	*-.01*	*4.4*	*-.2*
-30 * high	*-.13*	*.30*	*-2.9*	*6.7*
30-55 * high	*.16*	*.21*	*3.6*	*4.7*
+55 * high	*.37*	*.48*	*4.6*	*6.0*

This significance constitutes an external assessment of the significance of the first principal plane. As a matter of fact, it provides a method of determining the number of significant axes. It is clear that if the pattern observed on an axis is a mere artifact of the method, the positions of the supplementary categories on this axis should not be significant.

A sample of supplementary categories is displayed on figure 5.8. We note that educational level plays an important role along the vertical axis.

The category "low level of education" is positioned at about *15* standard deviations to the left of the origin along the first axis (test-value of *-15.3*), whereas "high level of education" is *10* standard deviations up the second axis (test-value of *+10.1*).

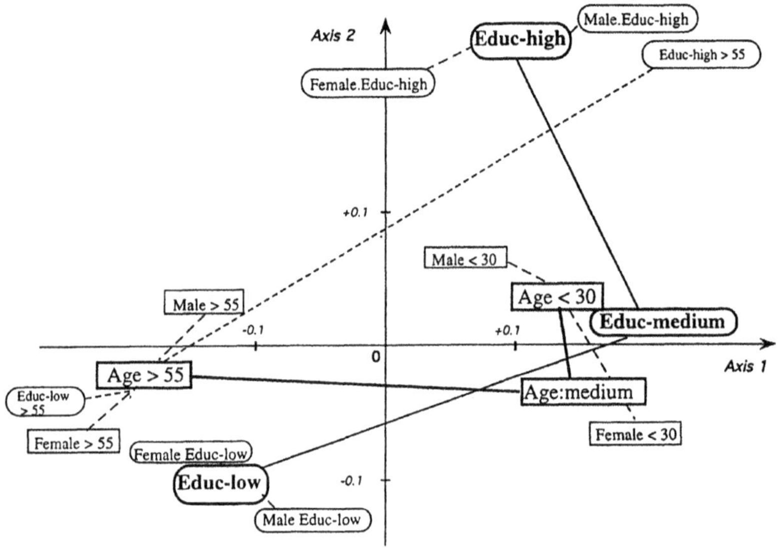

Figure 5.8
Positions of socio-demographic categories in the principal plane represented on figure 5.7

We find again that responses lacking in nouns characterize elderly people with a low level of education; that "abstract values" are mentioned mainly by educated people (see figure 5.2). With respect to the words used in the responses, the age category "over 55" appears to be extremely heterogeneous, since the educated people of this class (only 22 respondents) are situated on the upper right part of the display, quite far from the centroid of the whole category (lower left part).

The advantage of the direct analysis approach lies in its purely exploratory aspect. No *a priori* partition has been required to group the responses. Nonetheless, the results obtained prove very useful in practice, since they provide the user with a first overview of the main groupings of words and allow him/her to select possibly relevant criteria for agglomerating responses.

6

CHARACTERISTIC TEXTUAL UNITS, MODAL RESPONSES AND MODAL TEXTS

The typologies and visualizations of the preceding chapter produce global panoramas of lexical tables. These panoramas can be generated whether or not the data are aggregated, whether or not they are natural partitions of a body of work (chapters, works of one author, articles, discourses, etc.), whether or not they result from artificial groupings (open-ended responses organized by categories, documents organized by subject matter, etc.).

The spatial representations provided by correspondence analysis can be enhanced through the use of parameters of a more probabilistic nature. These parameters, called *characteristic elements* or *characteristic textual units*, are defined and illustrated in this chapter. They are used to describe groups of texts by displaying textual units that are present in a given group either a great deal more or a great deal less than in the overall corpus. These elements can be words, lemmas or segments (section 6.1).

In a more generalized, and therefore more complex fashion, it is also possible to exhibit, for any part of a corpus, characteristic associations consisting of collections of elements. Thus, in the case of open-ended questions, response groupings may be characterized by a few responses that are designated as *modal responses* (also called *characteristic responses*).

Briefly, modal responses generally contain a large number of characteristic words within a particular text. In the case of automatic information retrieval, modal responses might be modal documents (documents that characterize a particular theme). Thus the term modal response is used whether we are dealing with actual responses to open-ended questions or with documents (section 6.2).

In the case of longer documents, context elements of variable length (phrases, paragraphs, etc.) can be displayed.

6.1 CHARACTERISTIC ELEMENTS

The method presented below is an adaptation of classical statistical tests to textual material. The goal is to build a simple tool that can detect, within each of the parts of a corpus, which elements are used frequently as well as which elements tend to be used rarely. The hypergeometric model has been chosen because of its relative simplicity. Note that this model is not particularly analogous to text composition by a human being, a process which is far more complex: in this process, it is not possible to assume even the simplest of the hypotheses required within a hypergeometric model for performing statistical tests (independence of random samples, etc.). Consequently, the computed indices do not by any means represent directly interpretable probabilities[1]. Rather, they are of a purely descriptive nature. We shall use these indices to select characteristic terms (forms or segments) within each of the parts.

6.1.1 Computation of characteristic elements

We note the following quantities, expressed as occurrences of simple words (words can be either graphical forms [types] or lemmas) :

- k_{ij} - sub-frequency of word i in part j of corpus;
- $k_{i.}$ - frequency of word i in entire corpus;
- $k_{.j}$ - size (number of occurrences) of part number j;
- $k_{..}$ - size (number of occurrences) of corpus (or, simply k).

The probabilistic model

We start by imagining a *population* of objects with a total count k. Among all of these objects, we suppose that $k_{i.}$ objects are "marked", that is, that they possess a feature that differentiates them from the others: for example, balls of a particular color, or in our case, objects corresponding to occurrences (token) of a same word (type) whose total frequency is $k_{i.}$. The remaining objects of our population are all mixed together into a sub-total and considered as "unmarked". The number of unmarked objects is therefore equal to $k - k_{i.}$.

Now, using random sampling without replacement, we draw a *sample* that contains exactly $k_{.j}$ objects in our population. After this drawing, the number of marked objects contained in the sample is noted as k_{ij}.

The numbers k, $k_{i.}$, $k_{.j}$ defined above constitute the parameters of the model. In order to be able to make an assessment of the count k_{ij}, it must be compared to other similar counts among all possible samples comprised of $k_{.j}$ objects that can be drawn from the initial population.

---

1 For a presentation and some examples of the method of characteristic elements for texts, cf. for example: Lafon (1980).

For each sample of size $k_{.j}$, the value of the number k_{ij} of marked objects has to be between 0 and $k_{i.}$, i.e. the total number of marked objects. Conversely, again for a set of given numbers k, $k_{.j}$, $k_{i.}$, and for each number n whose value lies between 0 and $k_{i.}$, it is possible to count the number $N(n)$ of samples of length $k_{.j}$ for which k_{ij} is exactly equal to n.

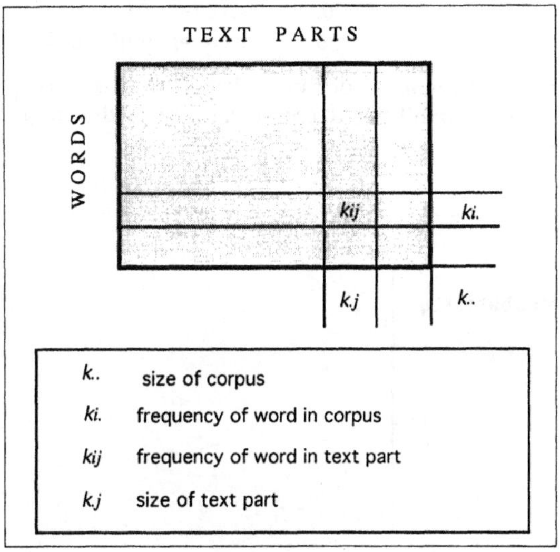

Figure 6.1
The four parameters for calculating characteristic elements

If each number $N(n)$ is divided by the total number of samples of length $k_{.j}$, the result is a probability distribution (with parameters k, $k_{i.}$, $k_{.j}$) of the set of numbers between 0 and $k_{i.}$.

The probability distribution for a sampling without replacement under the hypothesis of independence is the hypergeometric distribution. This distribution is similar to the multinomial distribution when the samples are relatively small compared to the population (then samplings with and without replacement are virtually the same)[1].

We note as: $Prob\ (k, k_{i.}, k_{.j}, n)$ the calculated probability of exactly n marked objects appearing in a sample without replacement of a sample of size $k_{.j}$

1 The classical hypergeometric distribution formula is:

$$Prob\ (k, k_{i.}, k_{.j}, n) = \frac{\binom{k_{i.}}{n}\binom{k-k_{i.}}{k_{.j}-n}}{\binom{k}{k_{.j}}}$$

among a population of total count k, knowing that the entire population has $k_{i.}$ marked objects

In figure 6.2 is shown an example of such a distribution that has been calculated with the parameters:

k	=	*160 000*	size of corpus
$k_{.j}$	=	*20 000*	size of part j
$k_{i.}$	=	*36*	frequency of word *i*.

It can be seen that the *mode* of this distribution (the most probable value) is equal to *4*. The probabilities decrease rapidly as the frequency varies away from this value.

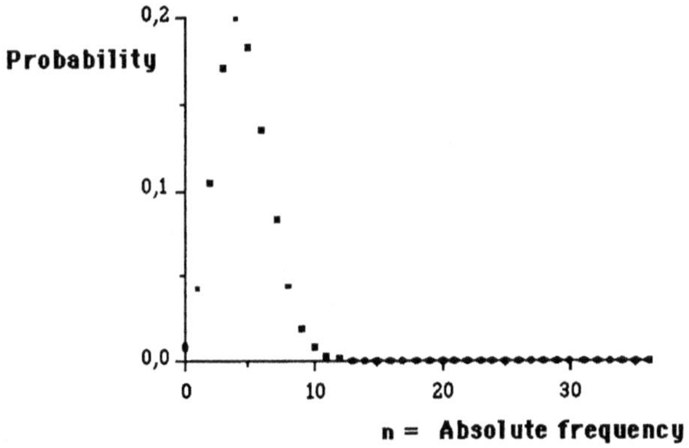

Figure 6.2
Hypergeometric distribution
for parameters k = 160 000, $k_{.j}$ = 20 000, $k_{i.}$ = 36

Now we can use the probability distribution thus built upon the parameters k, $k_{.j}$, and $k_{i.}$ to assess the absolute frequency n_0 observed when drawing our sample.

To this end we begin by placing n_0 in relation to the mode of the distribution. If the observed value n_0 is very close to the mode, there is not much we can say about the observed result.

If on the contrary it is much greater than the mode, we can calculate the quantity $P_{sup}(n_0)$. This quantity is the probability of having a number of marked objects greater than or equal to n_0 appear among the $k_{.j}$ objects drawn at random - still under the assumption that the hypotheses mentioned above hold true.

The probability $P_{sup}(n_0)$ is equal to the sum of all the probabilities for values of n that are greater than or equal to n_0. It is thus the sum of the probabilities *Prob* $(k, k_{i.}, k_{.j}, n)$ for values of n between n_0 and $k_{i.}$.

If n_0 is less than the mode, we calculate $P_{inf}(n_0)$ which is the probability of having a number of marked objects less than or equal to n_0 present - again assuming the same hypotheses.

This last probability is equal to the sum of the probabilities *Prob* $(k, k_{i.}, k_{.j}, n)$ for values of n between 0 and n_0.

If the probability $P_{inf}(n_0)$ is very small for a sample, the conclusion is that this sample has an "unusually" small number of marked objects compared to the whole set of samples of this type. Conversely, if the probability $P_{sup}(n_0)$ is very small for this sample, we conclude that the sample has an "unusually" large number of marked objects.

Practical computation of characteristic textual units

For each cell (i,j) of the matrix containing the number k_{ij} of occurrences of word i in part number j, we assess this sub-frequency with respect to the numbers k, $k_{i.}$, $k_{.j}$, using the model described above.

In order to select the probabilities that are to be considered as small, and consequently worthy of attention, a *probability threshold* is arbitrarily fixed at the beginning of the experiment. This threshold does not change during the calculations of the sub-frequencies contained in the table. This value is noted with the letter s.

Negative, positive, characteristic and banal textual units

If $P_{inf}(k_{ij})$ is smaller than the threshold, we conclude that the observed sub-frequency k_{ij} is relatively small, that is, word i is rather poorly represented in part j relative to what the hypergeometric model predicts. In this case we say that word i is a *negative characteristic word* for part j, and this is noted $S-$.

If on the contrary, $P_{sup}(k_{ij})$ is less than threshold s, we conclude that sub-frequency k_{ij} is relatively large, that is, that word i is relatively abundant in part j. We say that word i is a *positive characteristic word* for the part under consideration. This is noted $S+$.

It can also happen that neither of the two probabilities $P_{inf}(k_{ij})$ and $P_{sup}(k_{ij})$ is smaller than threshold s. In this case we say that word i is *banal* for part j. If a word never appears as a characteristic element (i.e. it is banal for each part of the corpus), this word is said to belong to the *basic vocabulary* of the corpus.

Diagnostics thus obtained by the method of characteristic elements in the table under analysis can then be re-ordered into more readable reports. The first report consists of lists in which the characteristic elements are sorted according to the words which they contain. Then each word in the corpus can be characterized in terms of its presence as either a positive or negative characteristic element in the different parts of the corpus.

For example, we can see below (table 6.1) the six diagnostics calculated on each of the sub-frequencies of the word *health* among the six categories of respondent of the corpus ***Life*** introduced in chapter 2.

Table 6.1

Frequency distribution of word *health*
(six age categories)

	Ag1	Ag2	Ag3	Ag4	Ag5	Ag6	*F*
word							
health	*47*	*100*	*114*	*111*	*95*	*144*	*611*
	[-E06]				[+E02]		

The column labelled *F* is the frequency of the term in the entire corpus. The characteristic element diagnostic (**ced**) consists of a sign and an exponent :

- +Exx refers to a *positive* characteristic element to which is associated a probability of the order *1/10* ^{xx}.
- -Exx refers to a *negative* characteristic element to which is associated probability of the order *1/10* ^{xx}.

Characteristic segments

Table 6.2 shows the frequency distribution of the occurrences of the segment *good health* in the ***Life*** corpus. It is a frequency distribution of occurrences of a special type: occurrences of the form *good* immediately followed by the form *health*. Characteristic elements can be computed for segments in a corpus.

Note that the diagnostics calculated from the repeated segments do not correspond exactly to the diagnostics calculated from the simple forms of which they are composed. Here again, a description by repeated segments enhances the description obtained from the simple forms.

6.1.2 List of characteristic units

Another type of report (see table 6.3) consists of ordering the characteristic elements within each of the parts of the corpus. This second type of report identifies the units (words or segments) that are over-represented as well as the units that are under-represented within each part. In other words this report provides a description of each part of the corpus by the terms that are used relatively frequently in that part, and by those that tend not to be used. In each of these reports, the characteristic textual units diagnostics are sorted in order of increasing probability, that is from the most characteristic to the most banal.

Table 6.2
Frequency distribution of words *good* and *health* and of segment *good health*

word	Ag1	Ag2	Ag3	Ag4	Ag5	Ag6	*F*
good	*39*	*45*	*47*	*45*	*44*	*83* [+E02]	*303*
health	*47* [-E06]	*100*	*114*	*111*	*95* [+E02]	*144*	*611*
good health	*9* [-E04]	*24*	*26*	*34*	*28*	*55* [+E03]	*176*

The problem of multiple comparisons

The simultaneous calculation of several characteristic element diagnostics runs into the obstacle of multiple comparisons, a problem familiar to statisticians. Suppose that the text parts are perfectly homogeneous and thus that the hypothesis of independence between the words and the parts is true. Under these conditions, on the average, out of *100* calculated characteristic element diagnostics, *5* are significant with respect to the probability threshold *5%*. In fact this *5%* threshold only makes sense for a single test, and not for multiple tests. In other words, the unsuspecting user will always find "something significant" at the *5%* level. A practical way to solve this difficulty is to choose a stricter threshold. In the context of analysis of variance, several procedures have been devised to overcome this difficulty [1]. However, in the present context, we are dealing only with orders of magnitude for the probability thresholds (the *ced* are expressed as powers of ten). As exemplified by the actual applications of chapter 7 (section 7.2.2 and 7.2.3) that involve probability levels varying from 10^{-4} to 10^{-35}, most practical situations lie outside the classical scope of multiple comparisons. These probabilities are much smaller than normally

1 As an example, the Bonferroni method, briefly, recommends dividing the probability threshold by the number of tests (number of comparisons in the case of the design of experiments). This reduction of the probability threshold is generally considered as too severe. Classical overviews and discussions about multiple comparisons are found in O'Neill and Wetherhill (1971), Mead (1988).

considered significant. The user can thus view these coefficients as an objective metric that has some informal interpretation as a probability.

Table 6.3

List of characteristic words and segments for the 6th category of the age partition

Word	*F*	*fp*	*ced*
I	286	90	+E04
worry	8	7	+E04
to	522	142	+E03
great	9	6	+E03
been	7	5	+E03
a little	6	5	+E03
good health	176	55	+E03
look	7	5	+E03
try	12	7	+E03
bit	11	7	+E03
grandchildren	30	13	+E03
money	171	26	-E03
home	90	10	-E03
employment	20	0	-E03
my work	23	0	-E03
my children	47	3	-E03
children	130	17	-E03
education	25	0	-E03
my job	37	1	-E04
my family	235	33	-E04
happiness	227	30	-E04
security	40	1	-E04
my	809	139	-E05
work	118	8	-E06
family	702	108	-E07
job	143	4	-E11

6.2 MODAL RESPONSES

A simple, straightforward technique addresses a number of objections that have been raised concerning the fragmentation inherent to any analysis that is limited to isolated words without placing them in their immediate context. This technique consists in the automatic selection of *modal responses* (also called *characteristic sentences* or *modal texts* in other types of applications). Modal responses are not manufactured responses that depict each text grouping, but actual responses given by the respondents, that are chosen because they are representative of a given category of individuals.

There are two ways in which modal responses are chosen: the first is based on calculations that make use of characteristic elements. The second is based on distance computations according to a simple geometric criterion (the chi-square distance).

6.2.1 Selection of modal responses using characteristic elements

A first way of selecting modal responses is the following: a modal response of a grouping is a response that contains, as much as possible, the most characteristic words of this grouping.

To each word i belonging to grouping j can be associated a *test-value* $t(i, j)$ that measures the deviation between the relative frequency of the word within group j relative to its global frequency calculated on the entire set of responses. This deviation is normalized in order to be able to be considered as a standardized normal variable under the hypothesis of random distribution of the word under consideration in the groups.

Practical computation of test-values

Within this framework, and using the notation defined in section 6.1.1, estimates of the expectation $m(i,j)$ and of the variance $s^2(i,j)$ of the hypergeometric variable k_{ij} observed in the cell (i,j), under the hypothesis of independence, are:

$$m(i,j) = \frac{k_{ij}}{k}, \qquad s^2(i,j) = k_{i.}\left(1 - \frac{k_{i.}}{k}\right)\left(\frac{k - k_{.j}}{k - 1}\right)$$

Thus, the quantity $t(i, j)$ such that:

$$t(i,j) = \frac{k_{ij} - m(i,j)}{s(i,j)}$$

is approximately a standardized normal variable under the hypothesis of random distribution of the word under consideration in the groups.

Under this hypothesis, the test value[1] $t(i, j)$ lies between the values -1.96 and $+1.96$ with a probability of 0.95. However since this calculation depends on a normal approximation of the hypergeometric distribution, it is only used when the counts within a grouping are not too small with respect to the total count [2].

---

1 Similar test-values have already been mentioned in chapter 4 in a different context, concerning the automatic description of the groups in a partition in table 4.5. See also section 5.3.3 of chapter 5.

2 For example, the order of magnitude of the parameters involved in the example of figure 6.2 is such that the Poisson approximation of the hypergeometric distribution is better than the normal approximation.

Table 6.4

List of characteristic words and corresponding test-values for four extreme categories of the age-education partition

		PERCENTAGE		FREQUENCY		Test-value
	WORDS	Within	Global	Within	Global	
L-30 = -30 * low						
1	keep	2.27	.48	4	50	2.33
2	not	2.27	.49	4	51	2.30
3	standard	1.70	.32	3	33	2.10
L+55 = +55 * low						
1	grandchildren	71	.29	19	30	4.19
2	I	3.46	2.38	92	248	3.99
3	as	1.43	.80	38	83	3.89
4	they	.49	.23	13	24	2.80
5	can	.64	.34	17	35	2.77
6	help	.41	.19	11	20	2.59
7	your	.41	.20	11	21	2.43

7	education	.04	.24	1	25	-2.52
6	money	1.05	1.63	28	170	-2.76
5	happiness	1.47	2.18	39	227	-2.96
4	security	.08	.38	2	40	-3.16
3	family	5.23	6.57	139	684	-3.27
2	work	.49	1.11	13	116	-3.73
1	job	.38	1.36	10	142	-5.61
H-30 = -30 * high						
1	friends	2.87	1.11	17	116	3.44
2	do	1.35	.45	8	47	2.60
3	want	1.01	.30	6	31	2.44
4	being	2.19	1.11	13	116	2.18
5	job	2.53	1.36	15	142	2.16
6	having	1.52	.67	9	70	2.11
7	things	.84	.27	5	28	2.06

2	wife	.00	.65	0	68	-2.10
1	health	2.70	5.85	16	609	-3.59
H+55 = +55 * high						
1	mind	2.55	.45	5	47	2.91
2	welfare	1.53	.21	3	22	2.42
3	peace	2.55	.74	5	77	2.17

Example of test-values

Table 6.4 presents the first seven positive characteristic elements and the last seven negative characteristic elements for the four "extreme" categories (corresponding to the two extreme age categories [less than 30 years and greater than *50*] and educational levels [low level and high level]). These *test-values* serve as a criterion for ranking words within each group. In the case of a medium sized corpus (such as a corpus of responses to open-ended questions) the rankings obtained are generally consistent with those given by the hypergeometric tests.

There are fewer than seven elements per group in table 6.4 when the test-value happens to be below the (practical) threshold of 2. In particular, no negative characteristic elements are displayed for the first and the last groups.

Characteristic elements are usually calculated on the corpus as a whole. However, Table 6.4, as the preceding chapter's correspondence analyses, is calculated on the contingency table partially listed in table 5.3, that is, on a table resulting from applying a frequency threshold (words appearing at least *16* times).

In this case, the choice of this frequency threshold for the words, which modifies the total number of occurrences of the partial text as well as the number of occurrences for each part of the text, has an influence on the calculated characteristic elements. In general, it is recommended that characteristic words be calculated on the entire corpus before carrying out selections through thresholds, a common practice in lexicometrics. However it can be seen that the results are robust as regards these computational options. When the goal is to complement a principal axes map that can only be carried out after a threshold is applied, then it is possible to work on the partial corpus, thus assuring compatibility of the lexical tables used.

Characteristic words and principal axes visualization

The results in table 6.4 can be compared to the proximities observed on figure 5.1 of chapter 5; it goes without saying that the results are more complementary than equivalent. The planar approximation of figure 5.1 causes a certain loss of information, but reveals an order and regular patterns that are not visible on table 6.4.

A useful representation consists in identifying in figure 5.1 the most characteristic points, and connecting them with the category-points they characterize. Anti-characteristic points (negative characteristic elements) are much more difficult to identify on maps.

The selection of modal responses

Table 6.4 is used as a reference for finding the "modal responses" (or characteristic responses) of each group. For each grouping, these words are ranked by their degree of significance: the smaller the rank of a word in this ranking, the more significant it is.

A simple empirical formula consists in associating with each response the mean rank of the words it contains: if this mean rank is small, it means that the response only contains words that are very characteristic of the grouping.

Even better than ranks, *test-values* can be used (the smallest mean rank then corresponds to the highest mean test-value). It is rather obvious that some choices have to be made: for instance, between a short response made up of a few very characteristic words, and a longer response, that necessarily contains some less characteristic words.

This calculation mode (criterion of characteristic words) has the property of favoring short responses, whereas the chi-square criterion tends to favor lengthy responses.

6.2.2 Selection of modal responses using chi-square distances

The principle of this selection is as follows: once the open-ended responses of k individuals (for example, let $k = 1000$) have been numerically coded by the computer, they give rise to a rectangular table **T** with k rows (individuals or responses) and as many columns as there are selected words ($V = 400$, for example).

Then a response is a row of the table, or a vector with *400* components. If this response is made up of *25* different words, only *25* of these components are non-zero.

A grouping of responses (an artificial discourse) is a set of row-vectors, and the mean lexical profile of this grouping is obtained by calculating the mean of the row-vectors of this set. If this grouping is based on the categories of a closed-end question whose responses are coded in a table **Z**, we have seen in section 5.1 that the aggregated lexical table **C** can be calculated with the equation:

$$\mathbf{C} = \mathbf{T'Z}$$

Thus we can calculate distances between responses and groupings of the same responses. Responses (rows of **T**) and groupings of responses (columns of **C**) are both represented by vectors in the V-dimensional space of words.

These distances express the deviation between the profile of a response and the mean profile of the group to which this response belongs. The distance that is used between these profiles of frequencies is the chi-square distance, thanks to its distributional properties stressed in chapter 3.

The distance between a row-point i of **T** and a column-point m of **C** is then given by the equation:

$$d^2(i,m) = \sum_{j=1}^{V} \frac{t_{..}}{t_{.j}} \left(\frac{t_{ij}}{t_{i.}} - \frac{c_{jm}}{c_{.m}} \right)^2$$

with the following notation:

- $t_{..}$ designates the overall sum of the elements of table **T**, that is, the total number of occurrences.
- $t_{.j}$ designates the sum of the elements of column j of **T** (number of occurrences of word j).
- $t_{i.}$ designates the sum of the elements of row i of **T** (length of response i).
- c_{jm} the *j-th* element of column m of **C** (number of occurrences of the word j in the grouping m).
- $c_{.m}$ is the sum of the elements of column m of **C** (total number of occurrences of grouping m).

For each grouping, these distances can be sorted in increasing order. Thus the most representative responses with respect to the lexical profile, that is, those whose distances are the smallest, can be chosen.

6.2.3 Implementation and examples

It is highly improbable that there should exist, among all of the original responses, a single response that summarizes by itself all of the properties of a category. Therefore, several characteristic responses are selected for each grouping, whatever the mode of calculation.

Modal responses keep to a minimum the impression of fragmentation that can result from looking at isolated words by situating characteristic words in their contexts.

For an open-ended question (here again we use the *Life* question as our example), and for a partition of the population (the partition into nine categories), we obtain, without any preliminary coding or other intervention:

- A visualization of proximities between words and categories (chapter 5).
- Characteristic elements or words for each category (section 6.1 of this chapter).
- Modal responses for each category.

Table 6.5
List of modal responses for four extreme categories of the age-education partition (Criterion 1: characteristic words)

Criterion 1 (average test-value)	Modal responses (4 per category)
Category 1	**Less than 30 years, low level of education**
.85 - 1	family, employment
.74 - 2	family work, money, have a good life
.70 - 3	keep a job, keep in clothing, transport, make sure family is all right
.54 - 4	that my children have a good education that I'm successful, that I can have my own home, to be able to afford some luxuries and all your needs, but not to have too much
Category 3	**Over 55 years, low level of education**
1.50 - 1	good health family, like to keep working for as long as I can
1.22 - 2	health, as long as we are healthy enough to live on comfortably, enough to eat
.92 - 3	not being miserable always like to help people if I can, looking after my health
.92 - 4	health, that my wife and I remain as active as possible, that we grow old together, that our children and their children are reasonably well and that we can carry on as happy as possible
Category 7	**Less than 30 years, high level of education**
1.33 - 1	friends, friends, my homelife
1.12 - 2	being content having enough money to do what you want to do, within reason, having good friends, having a fulfilling job to do, having some idea of what you want to do and the freedom to choose, protection of the environment
1.05 - 3	to have good friends around having a good job, living in a good area, having lots of freedom to do the things you want to do
.93 - 4	good living education, good job, money
Category 9	**Over 55 years, high level of education**
.97 - 1	togetherness, peace of mind, good health, religion, no
.64 - 2	not to die, peace of mind, don't like people living envious of each other
.63 - 3	peace of mind good health, happiness, enough money to keep a standard of living
.38 - 4	welfare of my family work, satisfaction, good health, travel

Table 6.6
List of modal responses for four extreme categories of the age-education partition
(Criterion 2: chi-square (χ^2) distance)

Criterion 2 (Chi-square distance)	**Modal responses (3 per category)**
Category 1	**Less than 30 years, low level of education**
.732 - 1	that my children have a good education that I'm successful, that I can have my own home, to be able to afford some luxuries and all your needs, but not to have too much
.738 - 2	to earn enough money to live to a decent standard, to own a nice house, to get on with the family and friends, to keep in employment
.812 - 3	to make money, and spend it with my young lady my flat-mates as friends, a good standard of living, a good car, just to be happy
Category 3	**Over 55 years, low level of education**
.818 - 1	I should say to carry on as we are from family point of view, everything smooth and no worries my daughters and sons are all right, grandchildren are all right, in health and family together, state of country, it seems to be in a good position at the moment, unemployment going down, I can't think of anything else
.838 - 2	my wife my grandchildren, my own children, to make my wife happy, to try and do as much good as I can in the world, to have good health, my family to have good health and always to keep myself occupied, one of the most important things, my garden
.842 - 3	good family life, having children and grandchildren visit often, as they do keeping happy, managing to keep going in our home, though we are both sick, making the most of each day as it comes
Category 7	**Less than 30 years, high level of education**
.758 - 1	to have good friends around having a good job, living in a good area, having lots of freedom to do the things you want to do
.818 - 2	being content having enough money to do what you want to do, within reason, having good friends, having a fulfilling job to do, having some idea of what you want to do and the freedom to choose, protection of the environment
.890 - 3	to break out of my conditioning to be clear and aware, to express my creativity as fully as possible, to go deep in love, to make beautiful things, to promote vegetarian eating and stop killing animals for food
Category 9	**Over 55 years, high level of education**
.774 - 1	work hard, try to make people happy, society in general, if I won a great deal of money, I'd love to help the unemployed and down and outs contentment in family, peace of mind
.837 - 2	my family a good welfare state for the less fortunate than myself, a good national health service and help for the aged,
.844 - 3	togetherness, peace of mind, good health, religion, no

Continuation of the example of chapter 5

First we show an example of modal responses corresponding to the four extreme categories of the partition into nine categories: less than 30 years and low level of education, less than 30 years and high level of education, over 55 years and low level of education, over 55 years and high level of education. Thus we simultaneously drop the intermediate age category and educational level.

For each of these four categories, we show the first four modal responses, for each of the two selection criteria. Tables 6.5 displays the responses selected according to the characteristic words criterion for each of the four previously mentioned categories.

The responses in table 6.5 are rather terse: since the word *friends* is the most characteristic word for category 7 and there is one response containing this word twice (see table 6.4 above), it is normal for it to be the most characteristic response when this criterion is used.

The corresponding selection criterion (1.33) is an arithmetic mean of the test-values corresponding to the words of this response. Only the words selected according to the frequency threshold are involved in this calculation. However, the initial response is printed in its entirety.

The almost caricatural nature of the responses selected by this first criterion is striking when tables 6.5 and 6.6 are compared. The reader should first consult the first responses of table 6.5, which constitute a summary, a semantic skeleton, that is fleshed out and completed by the responses chosen according to the chi-square criterion (table 6.6).

Comments

This example calls for several comments concerning the application as well as the method.

- The first comment is an elementary observation, but one that needs to be made at this point: a result that is reasonably coherent, a sort of summary of the initial data, has been obtained without any coding or preliminary interpretation of the text.
- The responses shown above are integral original responses, selected among 1043 recorded responses. They make it possible to restore words in their context, if there is one. In fact they could be positioned on figure 5.1, where each one would occupy the centroid of the words it contains. However this would clutter a representation that is already crowded.
- It can happen, although it is not the case here, that some rare forms appear among the characteristic responses. It must be recalled that the selection of responses is only done among words kept after applying a selection threshold to them according to their frequency (this global

frequency threshold is 16 in our example). But the listing of the responses respects their original wording. Thus it can happen, rather exceptionally, that a response contains only words whose frequency is less than the threshold, except for one single word that is a characteristic word for that category. With the lexical frequency criterion (but not with the chi-square criterion), this response might appear as a characteristic response, since the criterion coincides with the test-value of this single word, likely to be greater than the averaged test-values corresponding to the remaining responses.

We note that some modal responses are common to both criteria. For the first category of young persons with a low level of education, the fourth response according to criterion 1 (characteristic words) is the same as the first response resulting from the chi-square criterion. Similarly, for older educated people, the first modal response (criterion 1) is identical to the third according to criterion 2.

If we keep the first ten responses according to both criteria, we find many more common responses. These common responses can be considered to be the most reliable modal responses.

7

LONGITUDINAL PARTITIONS, TEXTUAL TIME SERIES

In many cases, a great deal of external information is available concerning texts compiled into a corpus: author, type, date (when texts by different authors are grouped into a corpus); socio-economic categories, age categories, educational level (in the case of aggregated open responses). This information can be used to establish a series of relationships (same author, similar or consecutive dates, same type, etc.).

This chapter introduces methods for working simultaneously with lexical analyses and *a priori* information for a corpus whose parts are consecutive in time: their purpose is not to uncover or confirm structures, but to test the reality of known structures, and to evaluate the level of dependence of the typologies obtained from the texts with respect to available *a priori* information.

7.1 LONGITUDINAL PARTITIONING OF A CORPUS

A longitudinal partition is obtained, for example, when open-ended responses are aggregated on the basis of values of a single numerical variable (age, monthly income or number of children) or on an ordinal variable, that is, a variable whose categories can be ordered in a way that corresponds to a degree of intensity for this variable (educational level, for instance).

A longitudinal partition is also obtained when the parts of a corpus of texts are arranged in chronological order (when they were written or published, for example). In the discussion that follows this type of partition is called a *longitudinal partition* of a corpus.

First we examine responses to an open-ended question grouped according to the categories of an ordinal variable (age categories). Then we proceed to the more general framework of textual time series by studying a series of political speeches.

7.1.1 Longitudinal partitioning example

In creating aggregates of responses of individuals belonging to each age category, our aim is to uncover variations associated with the variable "respondent age". To illustrate this type of partitioning, we use the ***Life*** corpus introduced in chapters 2 and 5.

Table 7.1

Responses to *Life* question grouped into 6 respondent age categories

code	*val.*	*occ.*	*words*	*hapax*	*fmax*	
Ag1	18-24	*2003*	*429*	*247*	*112*	my
Ag2	25-34	*2453*	*512*	*295*	*148*	my
Ag3	35-44	*2529*	*510*	*290*	*151*	family
Ag4	45-54	*2519*	*493*	*258*	*159*	my
Ag5	55-64	*1864*	*459*	*264*	*110*	my
Ag6	65-++	*3326*	*623*	*333*	*144*	health

Table 7.1 shows intervals used for creating 6 age groups (Ag1, Ag2, ..., Ag6). It can be seen that the resulting groups are comparable in size.

Cluster analysis of aggregates

Figure 7.1 shows the result of an agglomerative hierarchical cluster analysis of this partition (using the generalized Ward's criterion, as in the methods presented in chapter 4).

The dendrogram obtained from analyzing the table (6 age categories x 483 words of frequency greater than or equal to 3) has the property that its nodes consistently bring together consecutive age categories.

The extreme groups (Ag1 and Ag6) become agglomerated with the other groups at a very high level in the classification tree. This indicates a greater distinctiveness of these two groups with respect to the groups in between. As above, we interpret this finding by concluding that respondents belonging to neighboring age categories produce responses that are more similar among themselves than respondents belonging to more distant age categories.

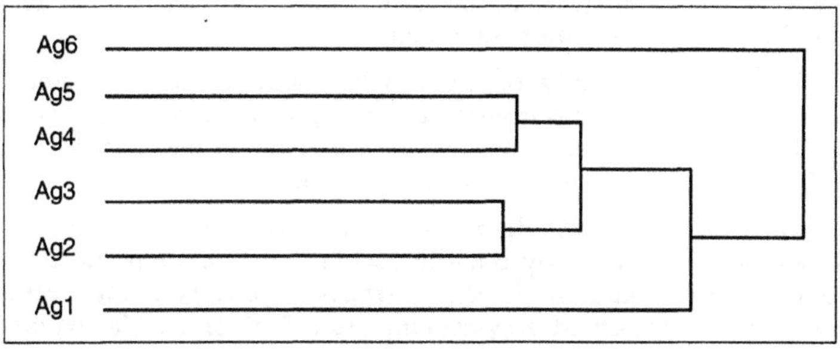

Figure 7.1
Agglomerative hierarchical cluster analysis based on partitioning of responses to *Life* question into *6* age groups.

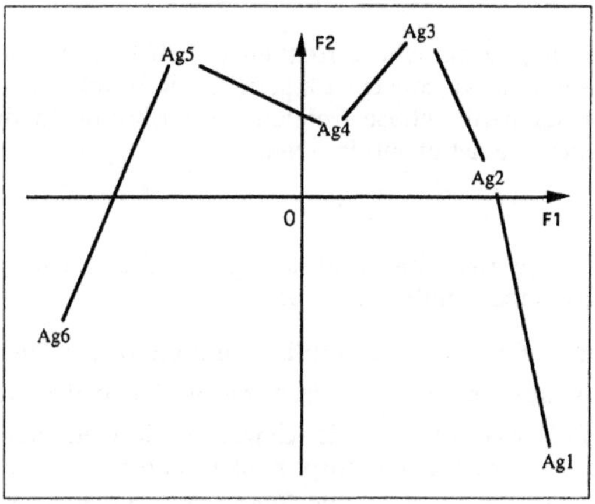

Figure 7.2
Plane spanned by the first two dimensions.
Table cross-tabulating *6* age categories and *483* words
(*Life* open question)

7.1.2 Analysis of *age category* gradation

When we perform a correspondence analysis - figure 7.2 - on a table of distances calculated on the lexical information belonging to each part of the *Life* corpus, the points corresponding to age categories (Ag1, ..., Ag6) describe a *horseshoe pattern*. Inspection of the principal axes suggests that the analysis is dominated by the presence of a gradation. The words and segments that have an extreme principal coordinate correspond to terms used by the youngest or oldest respondents.

These results are typically obtained when a table characterized by a gradation among aggregates is analyzed through correspondence analysis. If the gradation is smooth, the graph of the elements on the planes generated by the first principal axes has a characteristic horseshoe shape[1].

Correspondence analysis displays in a rather complex fashion a phenomenon that is basically one-dimensional. In the statistical literature this phenomenon, known as the *Guttman effect*, is named after the statistician who performed a systematic study[2] of data tables having this type of structure. However, more specific methods are needed to uncover which terms are quantitatively associated with the overall gradation observed on the first planes. These methods are presented below.

7.1.3 Adjacent characteristic elements

In order to identify terms whose frequency distribution is associated with the overall gradation, we construct a series of indicators. Let us consider a term (word or segment) whose frequency distribution in the *n* parts of a corpus is given by the list of sub-frequencies:

$$f_1, f_2, f_3, \ldots, f_n$$

As usual T designates the total length of the corpus and the list: $t_1, t_2, t_3, \ldots, t_n$, gives the length of each of the parts.

We saw in chapter 6 that these quantities are used to calculate a diagnostic S_j for each of the sub-frequencies f_j. From these diagnostics are selected *first level characteristic elements*, that is, characteristic elements having to do with elementary periods of the corpus taken one at a time. In the case of first level characteristic elements, the comparison is made, for a given part j and a given word i, on four numbers (T, t_j, F, f_i) where F is the frequency of the term in entire corpus.

We designate as *second level adjacent characteristic elements* (cf. figure 7.3), characteristic elements calculated in the same fashion for each pair of consecutive parts.

For the *second level*, diagnostics are calculated from the numbers T, $tj+tj_{+1}$, F, $f_i + f_{i+1}$. Thus there are $n - 1$ second level diagnostics.

---

1 Starting with the second axis, the axes are functions of the first axis (a quadratic function for axis 2, a cubic function for axis 3, etc.).

2 Cf. Guttman (1941, 1946 and 1950), Benzécri (1973) and Van Rijckevorsel (1987).

We define in the same way diagnostics of level k, that is, the $n - k + 1$ diagnostics calculated for sequences of k consecutive parts. The highest level groups together the two diagnostics of level $n - 1$.

When they are calculated in this manner, adjacent characteristic elements can be used to identify some of the irregularities in the frequency distribution of a term within the parts of a corpus.

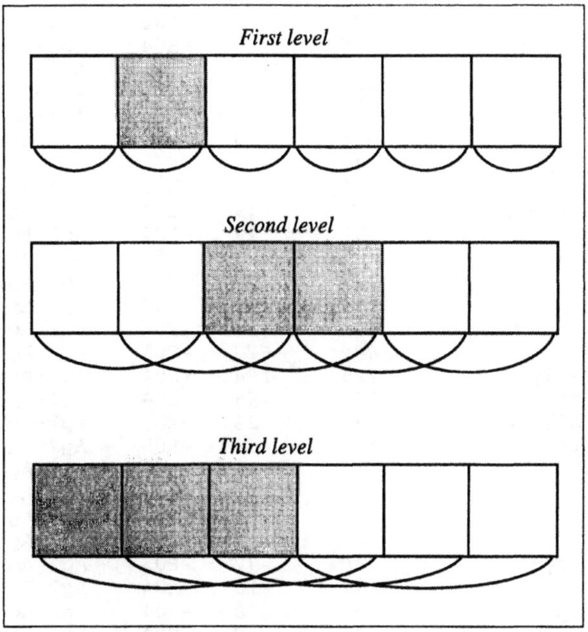

Figure 7.3

Computation of adjacent characteristic elements (examples of three levels)

Application to Life question

The characteristic element diagnostics shown in table 7.2 are concerned with terms whose frequency is greater than 5 occurrences in the entire ***Life*** corpus.

For each term we began by calculating the first level diagnostics relative to the frequency distribution of the term in each of the parts of the corpus. In a second phase, we calculated the diagnostics that concern consecutive pairs of parts (level 2, 3, etc.).

Table 7.2

Major adjacent characteristic elements for corpus of *Life* responses grouped into 6 respondent age categories

term	F	pf	ced	group
to do	*25*	*13*	+E06	Ag1-
being	*116*	*32*	+E05	Ag1-
friends	*116*	*32*	+E05	Ag1-
good job	*13*	*8*	+E05	Ag1-
having	*70*	*20*	+E04	Ag1-
a good job	*6*	*5*	+E04	Ag1-
a good	*54*	*18*	+E04	Ag1-
a	*300*	*62*	+E04	Ag1-
money	*170*	*80*	+E06	Ag1-Ag2
career	*11*	*9*	+E04	Ag1-Ag2
be happy	*41*	*23*	+E04	Ag1-Ag2
to get a	*6*	*6*	+E04	Ag1-Ag2
being happy	*30*	*19*	+E04	Ag1-Ag2
my job	*37*	*29*	+E04	Ag1-Ag3
education	*25*	*20*	+E04	Ag1-Ag3
job	*143*	*128*	+E12	Ag1-Ag4
work	*118*	*100*	+E07	Ag1-Ag4
family	*686*	*582*	+E07	Ag1-Ag5
my	*809*	*670*	+E05	Ag1-Ag5
my family	*225*	*194*	+E04	Ag1-Ag5
happiness	*227*	*198*	+E04	Ag1-Ag5
the children	*25*	*21*	+E07	Ag2-Ag3
children	*125*	*66*	+E05	Ag2-Ag3
wealth	*11*	*10*	+E04	Ag2-Ag3
of	*312*	*193*	+E05	Ag2-Ag4
can	*35*	*8*	-E04	Ag2-Ag4
do	*47*	*17*	-E05	Ag2-Ag5
security	*40*	*37*	+E05	Ag2-Ag5
want	*31*	*10*	-E04	Ag2-Ag5
to keep	*27*	*8*	-E04	Ag2-Ag5
keep	*50*	*19*	-E04	Ag2-Ag5
health	*611*	*564*	+E06	Ag2-Ag6
welfare	*22*	*11*	+E04	Ag3-
welfare of	*13*	*8*	+E04	Ag3-
the	*331*	*155*	+E06	Ag3-Ag4
the family	*78*	*47*	+E06	Ag3-Ag4
need	*10*	*9*	+E04	Ag3-Ag4
to	*522*	*203*	-E05	Ag3-Ag5
are	*65*	*57*	+E04	Ag3-Ag6
me	*33*	*31*	+E04	Ag3-Ag6
be	*137*	*10*	-E04	Ag4-
grandchildren	*30*	*30*	+E09	Ag4-Ag6
I	*248*	*158*	+E04	Ag4-Ag6
they	*24*	*21*	+E04	Ag4-Ag6
religion	*13*	*13*	+E04	Ag4-Ag6
good health	*176*	*117*	+E04	Ag4-Ag6
as	*83*	*44*	+E04	Ag5-Ag6
worry	*8*	*7*	+E04	Ag6-

At the end of this process, the diagnostic with the smallest associated probability is shown for each term.

> ***Guide to reading table 7.2***
>
> The characteristic elements have been selected on terms (words or segments) whose frequency is greater than *5* occurrences in the corpus. Only the most significant diagnostics are shown in this table (probability index less than $1/10^4$).
>
> The column labelled ***F*** is the frequency of the term in the entire corpus.
>
> The column labelled ***pf*** is the partial frequency of the term in the part or group of consecutive parts for which the characteristic element diagnostic was computed.
>
> The characteristic element diagnostic (column labelled **ced**) consists of a sign and an exponent (+Exx refers to a positive characteristic element to which is associated a probability of the order $1/10^{xx}$).
>
> The part, or the group of consecutive parts, of the corpus for which the positive characteristic elements have been determined appears in the last column.
>
> The adjacent characteristic elements are sorted from youngest to oldest.

The characteristic element diagnostics presented in table *7.2* correspond to distributions that are the most remarkable in their distortion compared to a distribution resulting from a random drawing. The longitudinal characteristic elements thus calculated are a mixture of easily predictable observations and unexpected discoveries. The words *job* and *work* are close with respect to their distribution in the age categories (somewhat more frequent in categories *1* to *4*). *Good job, career* and *money* and also *friends* and *being happy* tend to be mentioned by the younger age categories (Ag1 and Ag2). *Health* is most often mentioned by older age categories. The substantial presence of the article *the* in the responses of age categories Ag3 and Ag4 can be explained by the fact that these highly educated respondents tend to formulate longer sentences with a more precise syntax, and with a heavier usage of nouns (cf. chapter *5*, the notes following section *5.5*).

7.2 TEXTUAL TIME SERIES

In the field of text analysis, and particularly in the socio-political field, analyses of the press and historical studies often involve building a type of corpus by sampling a single source of text over time. The period of time can be of lesser or greater length. Texts thus assembled into corpora are often drawn up under very similar conditions, sometimes by the same author. Their temporal spacing is a reason for performing comparative analyses and ascertaining what time-related variations might be present.

We give the name *textual time series* to those homogeneous corpora comprising texts generated under similar conditions (if possible by a single author) and presenting comparable lexicometric characteristics. An analysis based on the time dimension might be fruitful. For example, one might decide to undertake an analysis of the evolution of the vocabulary of an author during the entire period of creation of his/her works. This analysis might initially be based on the assumed date of writing each work, even though this criterion might be modified later.

Analyses performed on textual time series have brought out the importance of the phenomenon of the overall evolution of vocabulary with time. Any author producing texts over a long enough period of time constantly uses new vocabulary words that eventually replace, in terms of frequency, other words whose usage becomes rarer. In most cases, this vocabulary renewal is the fundamental lexicometric characteristic of a time series.

It follows that vocabularies of parts corresponding to consecutive periods of time are generally more similar among themselves than those separated by longer periods of time. Of course, it can happen that more extreme lexical disruptions overwhelm this chronological evolution.

7.2.1 *Speeches* time series

The *Speeches* time series presented in chapter *2* is a collection of *68* television speeches of F. Mitterrand between July 1981 and March 1988. This corpus is an excellent example of a *textual time series.*

Partitioning into 7 years

A chronological partition of the corpus is obtained by grouping the speeches into *7* parts where each part corresponds to a period of one year (from May *21* to May *20* of the following year).

The main lexicometric characteristics of the *7* parts are found in table *7.3.* As shown in this table *7.3*, this partitioning produces a relatively balanced division of the corpus from the point of view of number of occurrences within each part and total number of words. Selection of *1 397* words of frequency greater than *20* occurrences represents *256 855* occurrences, or *86%* of the total number of occurrences in the corpus.

The results of the correspondence analysis of the table (*1397* words × *7* periods) with respect to the *7* periods are summarized in figure *7.4*. An inspection of the eigenvalues relative to each axis shows that there are three prominent axes ($\lambda_1=0.026, \tau_1=29.0\%, \lambda_2=0.017, \tau_2=18.9\%, \lambda_3=0.014, \tau_3=16.2\%$).

Table 7.3

The 7 periods of *Speeches* corpus

	code	*years*	*occurr.*	*words*	*hapax*	*fmax*
1	Mit1	(81-82)	*30867*	*4296*	*2298*	*1388*
2	Mit2	(82-83)	*41847*	*5077*	*2611*	*1807*
3	Mit3	(83-84)	*53050*	*5622*	*2818*	*2040*
4	Mit4	(84-85)	*36453*	*4601*	*2362*	*1285*
5	Mit5	(85-86)	*53106*	*5509*	*2726*	*1975*
6	Mit6	(86-87)	*41205*	*4920*	*2499*	*1523*
7	Mit7	(87-88)	*40730*	*4748*	*2417*	*1526*

We immediately see that the positioning of the periods along the first principal axes mostly follows the lexical time series scheme outlined above. The most notable deviation from this scheme is the position of the last period along the second axis. Period Mit7 occupies a more central position than one that would correspond to a homogeneous lexical evolution over time.

Taking into account the size of the contingency table and its overall count *(256 855)*, it is obvious from a statistical point of view that the particular location of the period Mit7 is highly significant. Based on this observation, an analysis of the hypothesis that there occurred a perturbation to the presidential speeches during the last year of the seven-year term is in order.

Note that the evolution is also far from homogeneous from one period to the next. Thus periods Mit1 and Mit2 are close on the display of the first two dimensions whereas periods Mit5 and Mit6 are distant. Periods Mit3 and Mit4 are noteworthy in that they disrupt the alignment of the Mit1-Mit6 series of points.

7.2.2 Chronological characteristic elements

Analyses of chronological characteristic elements (longitudinal characteristic elements in the case of a time series corpus) can be carried out to uncover terms that are used particularly often (as well as particularly rarely) during a single time period or a group of consecutive time periods.

Table *7.4* shows a list of the major chronological characteristic elements of the *Speeches* corpus. Only a very few words that are significant from a statistical point of view[1] have been selected for this table. The list begins with a characteristic element diagnostic on the word *we* whose relative abundance in the first parts is detected.

Note that, on the contrary, the frequency of the first person singular word "*I*" is smaller in the early periods and tends to increase with time.

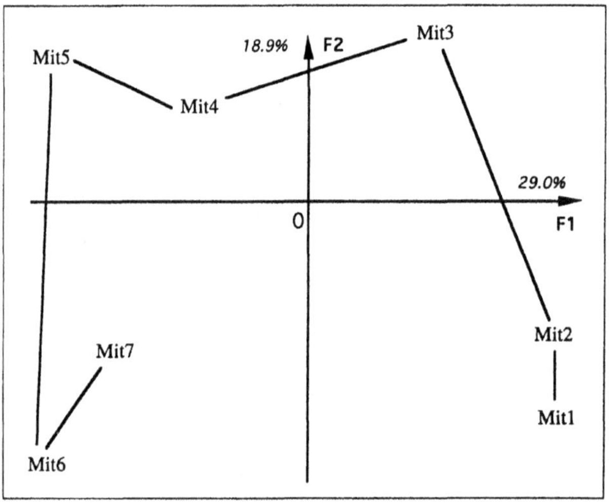

Figure 7.4

***Speeches* corpus : sketch of first two dimensions of a correspondence analysis of table [*1397* words (F ≥*20*) ×*20* periods]**

Period *1* is characterized by the presence of the terms *nationalizations, reforms, boost*. We observe that these expressions disappear in ensuing periods.

A more detailed inspection of this list can uncover themes relative to each period and their fluctuations with time. As usual, reports based on sorted probabilities serve to highlight deviations.

As above, these lists can be sorted by periods if the goal is to illustrate fluctuations of vocabulary with time. The volume of these reports can be controlled by varying the frequency thresholds involved in the processing : frequency of terms and probability thresholds.

1 The probability index for these forms is less than $1/10^{14}$.

Table 7.4

Major chronological characteristic elements of *Speeches* corpus (cf. guide for reading table *7.2*)

term	F	pf	ced	group
we	2059	1148	+E35	01-03
Iran	50	41	+E27	07-
nationalizations	42	31	+E22	01-
students	28	27	+E22	06-
channel	39	34	+E21	05-
Israel	71	54	+E20	01-02
majority	212	130	+E19	05-06
reforms	39	27	+E18	01-
has	2863	1875	+E18	04-07
boost	32	24	+E17	01-
have	523	458	+E17	01-05
of	11544	3195	+E17	01-02
it	3240	2643	+E17	03-07
I do not	794	567	+E16	04-07
we have	413	366	+E16	01-05
percent	204	193	+E16	01-05
that	1636	1096	+E15	04-07
not	4067	2092	+E15	05-07
country	748	358	-E15	03-06
fifth	35	27	+E14	05-
I	5024	3156	+E14	04-07

7.2.3 Characteristic increments

In order to complete the analysis of a vocabulary's evolution over time, there remains the task of constructing tools that can be used to signal sudden changes in a term's usage during a given period compared to preceding periods.

Calculation of characteristic increments

In what follows, we continue to work with a lexical table of the same type as those we analyzed previously. However the frequency k_{ij} of word i in part j is contrasted with quantities that are different from those used thus far for calculating simple characteristic elements.

If we establish a period j (j varies between 2 and p, the total number of periods in the corpus, as we are dealing with increments), it is possible to compare the frequency k_{ij} of a term in a period with its frequency in the overall set of parts from *1* to *j*.

To accomplish this comparison we begin by calculating the four parameters of the hypergeometric model. Two of these parameters are identical with respect to calculating simple characteristic diagnostics. These are the parameters:

k_{ij} - frequency of term *i* in part *j*;
t_j - number of occurrences in part *j*.

The two others are analogous to parameters used in the simple characteristic element calculations for which the summation is now performed not on the entire set of parts but on the first *j* parts (figure 7.5).

T^j - number of occurrences of corpus reduced to only the first *j* parts of the corpus;
F^j_i - frequency of term *i* in the same sub-corpus.

These last two parameters have superscripts because the summation is a partial summation limited to the first periods of the corpus.

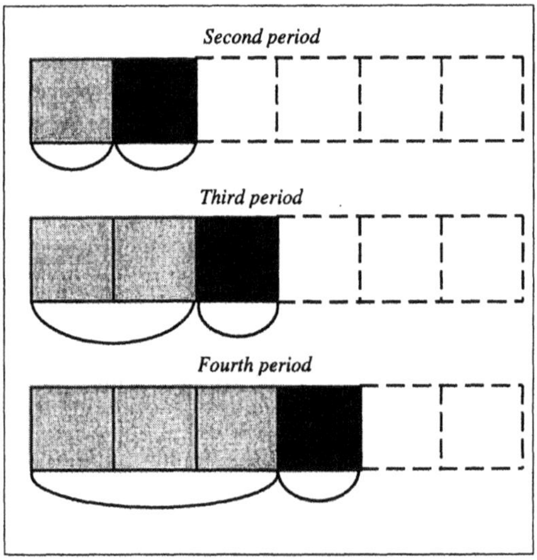

Figure 7.5
Computation of characteristic increments (schematisation of the first periods)

Table 7.5

Major characteristic increments in the *7-th* part of *Speeches* corpus (period: *4/12/87 - 3/4/88*)

term	F_7	f	incr.
Iran	53	41	↑E25
Arab	34	23	↑E13
Arab world	21	17	↑E12
of education	20	16	↑E11
education	23	17	↑E11
investigation	23	17	↑E11
Iraq	32	18	↑E08
election	35	18	↑E07
president	303	73	↑E07
of weapons	27	15	↑E07
a president	28	15	↑E07
political	105	34	↑E07
weapons	93	32	↑E07
judge	35	17	↑E07
country	748	151	↑E07
way	25	13	↑E06
yes	256	63	↑E06
candidate	26	14	↑E06
there	1749	305	↑E06
really	143	40	↑E06
there is (there are)	1014	191	↑E06
disarmament	38	17	↑E06
------------	------	------	------
we have	413	27	↓E06
inflation	83	0	↓E06
have	523	35	↓E07
youth	134	2	↓E07
we	2059	182	↓E12

Note that for the last part of the corpus, and the last part only:

$$T^p = T; \; F^p{}_i = F_i$$

In other words, for the last part the calculation of characteristic increments gives identical results to the calculation of simple characteristic diagnostics. For each given period, we can thus obtain a list of terms whose usage has significantly increased as well as a list of terms that have suddenly disappeared.

An analysis of the major characteristic increments of a corpus can be conducted to pinpoint the most abrupt changes throughout the corpus. In table *7.5* we display the most important diagnostics concerning part Mit7, which, as we saw on the first plane resulting from the correspondence analysis of the table (*1397* words × *7* parts) occupies an unusual position with respect to the evolution of the whole set.

Guide to reading table 7.5

Column F_X is the frequency of a term in the first x parts of the corpus. (The last period is a special case where $F_X = F$, total frequency of the word in the corpus).
Column f is the partial frequency of a term in the part for which the characteristic increment diagnostic has been computed (the last part in this case).

Exponent of characteristic increment diagnostic:

- ↑Exx refers to a positive characteristic *increment* [1] or over-use of a characteristic element relative to the set including the part and all the *preceding parts*. To such increment is associated a probability of order $1/10^{xx}$.
- ↓Exx refers to a negative characteristic *increment* or under-use of a characteristic element relative to the set including the part and all the *preceding parts*, to which is associated a probability of order $1/10^{xx}$.

The diagnostics shown are sorted first on positive or negative characteristic elements and then on the probability level.

The characteristic increments are calculated on sub-frequencies of terms (words or segments) with frequencies greater than *20* occurrences in the corpus. Only diagnostics whose probability level is less than $1/10^6$ are shown.

This table very clearly illustrates the emergence, in the course of the 1987-88 presidential year, of a vocabulary linked to international conflicts of the day (*Iran, Arab world, Irak, weapons*) and to legal affairs associated with terrorist attacks (*judge, investigation*), whereas preoccupations having to do with *inflation* and *youth* are set aside to a dramatic extent. Note also the negative characteristic increments of forms involving the first person plural (*we, we have*).

[1] The previous notations +Exx and -Exx referred to characteristic elements of a group with respect to the whole corpus.

In displaying just a few words associated with the most exceptional diagnostics of characteristic element increments, we are purposely limiting ourselves to an illustration of the method. It goes without saying that a more extensive selection of terms (easily obtained by choosing a less drastic probability threshold) would uncover increments whose existence might be less obvious to a political analyst.

7.2.4 Parallel analysis of a lemmatized corpus

Counts of graphical forms on the one hand and counts of lemmatized units on the other can lead to different results with respect to the principal lexicometric characteristics of a corpus [1]: number of words, hapaxes, maximum frequency, etc. The purpose of this section is to evaluate the impact of lemmatization on results of multivariate analyses presented in the preceding chapter. For this comparison, performed on counts of occurrences of each lemma in the *7* periods of the corpus, we use the same selection thresholds established for the analysis carried out on the graphical forms. Then we analyze a table crossing the *1312* lemmas whose frequency is at least *20* occurrences, and the *7* periods of the experiment discussed previously.

The results of the correspondence analysis of this table are summarized for the *7* periods in figure *7.6*. The symbols designating the *7* periods of the corpus (Mit1.L, Mit2.L,...,Mit7.L) include the letter *"L"* to recall that the counts are based on lemmas.

Again three axes are prominent, and their eigenvalues and percentages of variance are very close to those obtained on the words analysis ($\lambda_1=0.024$, $\tau_1=28.7\%$, $\lambda_2=0.016$, $\tau_2=19.4\%$, $\lambda_3=0.014$, $\tau_3=16.6\%$).

An inspection of the relative positioning of the periods on the plane spanned by the first two dimensions confirms the initial impression that the results are very similar to those obtained on the graphical forms (cf. fig. *7.4*). A comparison of the list of the most strongly contributing units for each of the types of counts (lemmas and graphical forms) that are used as active elements in each of the analyses shows again that the differences are minor overall.

Similarly, the lists of chronological characteristic diagnostics and strongest characteristic increments calculated on the basis of the lemmatized units reflect the diagnostics obtained on the graphical forms.

Thus, using tables built from very different counts, statistical methods such as correspondence analysis, hierarchical clustering, chronological characteristic elements, produce very similar results. The experiment presented above is by no means a definitive proof that contrasts uncovered by multivariate analyses among texts are independent of the choice of units

[1] Cf. the discussion of the ***Speeches*** time series in section 2.7.

to count. However this has been a common observation when analyzing large corpora.

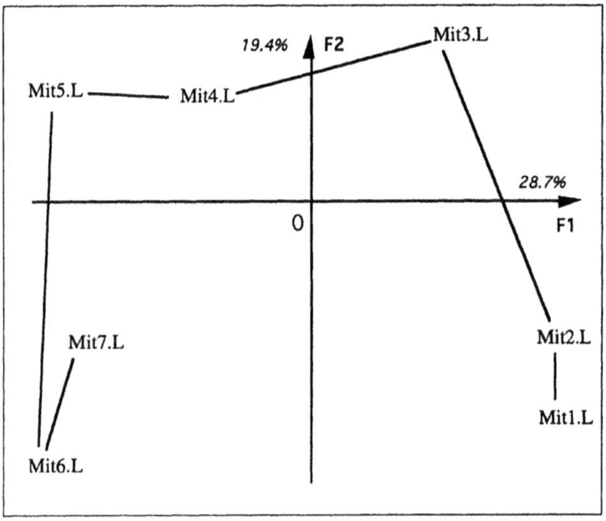

Figure 7.6

***Lemmatized Speeches* corpus**
Two first dimensions of a correspondence analysis of table
[*1312* lemmas (F ≥ *20*) × *7* periods]

In an absolute sense it is always preferable to have a double network of counts (graphical forms and lemmas). However, a *complete lemmatization* on a large corpus is a costly operation. Whereas it is indispensable in research work, its justification is much weaker if the goal is to obtain rapid visualizations or typologies of corpus parts with a certain amount of lexical richness. In this latter case one can work on graphical forms (or words) without worrying about bypassing essential information.

On the other hand, if the analysis involves short, ungrouped texts, lemmatizing texts offers a point of view that is sometimes different, and often complementary (see for instance the examples of sections 5.1.8 and 5.3.3 relating to responses to open-ended questions).

8

TEXTUAL DISCRIMINANT ANALYSIS

The statistical methods discussed in the previous chapters are applicable mainly in the exploratory phase (also known as the descriptive phase) of an analysis. However, data *exploration* is more dynamic and interactive than simple data description. It uses multivariate statistics to obtain visualizations or groupings of elements that can be either whole texts or items within texts. It looks for associations and structures as well as interesting summaries.

These methods complement a broad spectrum of techniques devoted to a type of research often called decision research, even though they are actually decision aids. In textual statistics decision aids are used for attributing a text to an author or a period, choosing a document within a data base in response to a request, and coding information expressed in everyday language.

The statistical procedures used for these attributions, choices or coding are based on discriminant analysis. Their aim is to predict the assignment of an "individual" to a group or a category (among several possible groups/categories) on the basis of variables measured on this individual[1]. This prediction is made possible thanks to a learning phase carried out on a set of individuals for whom both variables and groups are known (*learning* or *training sample*). However, the concepts of variable and individual are less appropriate for researchers who are analyzing textual materials than for biometricians or demographers. The definition of statistical units constitute, as we have seen in section 2.1, a crucial phase of research. Textual

[1] Early discriminant analyses were carried out on biometric and anthropometric measures by statisticians Fisher (1936) and Mahalanobis (1936), who were attempting to predict belonging to ethnic groups on the basis of measurements of the skeleton. They were the first to use the technique that is sometimes known as linear discriminant analysis: it is one of the oldest methods, and it is also one of the methods that is most commonly used today.

discriminant analysis puts a great deal of emphasis on these preparatory stages.

In section 8.1 we present two major areas of concern that are related to the statistical procedures of discriminant analysis and pattern recognition in textual data analysis. The first area deals essentially with the form of the textual data and disregards the content. It is based on stylometry and is developed in sections 8.2 (a brief review of the units and indices of stylometry) and 8.3 (an example of a statistical model in stylometry, and a recommendation for alternative methods). The second area, designated as *global discriminant analysis*, takes content of textual data into account, and sometimes form as well. Global discriminant analysis is presented in section 8.4, and developed by means of an example in section 8.5.

8.1 TWO MAJOR AREAS OF CONCERN IN TEXTUAL ANALYSIS

Among the most common applications of textual analysis techniques related to discriminant analysis, there are two main areas:

- Applications to literary corpora (for example, attribution to authors or dating) attempt to go beyond a text's content in order to capture the characteristics of its *form* (often equivalent to its style) on the basis of statistical distributions of vocabulary, and indices and ratios. The idea is to capture the "invariant" aspects of an author or of an era, that may be hidden to the ordinary reader.
- At the other extreme, applications in Information Retrieval, automatic coding and analysis of open-ended responses, essentially deal with *content*, the meaning and the substance of texts. The way in which a response is formulated is less important than the fact that it belongs to a group of documents that are relatively homogeneous as to their content. This second family of applications is referred to as *global discriminant analysis*.

In practice, as we mentioned before, it is not always practical to maintain this distinction. Again, in the process of statistically analyzing open-ended questions, the patterns of responses, the connotations carried by apparently equivalent terms constitute a wealth of information that often makes it possible to associate nuances to the more obvious content of a text.

8.1.1 Discriminant analysis based on patterns: stylometry

In numerous applications to literary, political and historical corpora, discriminant analysis has been used to attribute texts of dubious origin (poems, plays, novels, religious or sacred texts) to known authors, or to establish the era in which they were written. These attribution procedures have not always used the biometricians' discriminant analysis, as is seen in the pioneering work of Yule (1944).

In his book *Statistics and Truth* C.R. Rao (1989) mentions more recent work done by Thisted and Efron (1987) to attribute a poem discovered in 1985 to Shakespeare. Later we shall discuss this work at greater length. He also mentions Mosteller and Wallace's impressive work (1964) on attributing some of the Federalist Papers.

The *Federalist Papers* constitute 77 political texts published anonymously at the end of the eighteenth century. Among these, 12 texts were not able to be attributed to either of two possible authors. The statistical analysis identifies the most probable author of these 12 controversial texts. More recent work has been performed by Holmes and Forsyth (1995) on *The Federalist Papers*, and, in a similar vein, by Holmes (1992) on the homogeneity of the Mormon Scriptures. This author has also written a general overview of the methods of *stylometry* (1985)[1].

Most of these methods use numerical indices constructed on the basis of word lengths, sentence lengths, function word frequencies, richness of vocabulary and frequency distributions of words. A recent and thorough survey of authorship attribution is given by Holmes (1994).

The systematic use of correspondence analysis and of clustering techniques (cf. Benzécri et al., 1973; 1981) has provided new directions of research.

8.1.2 Global discriminant analysis: information retrieval, coding, validation

Today, Information Retrieval has become an autonomous discipline with its own publications, conferences, software, terminology and concepts (cf. Salton and McGill, 1983; Salton, 1988).

This specialization owes its existence to the size and context of the problems encountered. Generally very large and sparse matrices are being dealt with. These matrices come from specialized vocabulary and key-word counts, depending on the subject area, within bodies of work that include several hundred thousand documents that are often short and sometimes stereotyped. This operation is often very pragmatic. The idea is to have a tool that works, with established failure rates and predefined systems of costs and constraints.

In this context, there are several outside sources of information that can be called upon to resolve classification problems: syntactic analyzers, preliminary steps toward gaining an *understanding* of the search,

---

1 The second version of the work of Mosteller and Wallace (1984) also contains a general panorama of attempts at authorship attribution.

dictionaries or semantic networks to lemmatize and eliminate ambiguities within the search, and possibly artificial corpora that resort to experts[1].

However, many of these techniques, and among them the most efficient ones - according to their authors - use multivariate tools that are very similar to those advocated by Benzécri (1973, 1977, 1981) or by Lebart (1982a) in the case of large, sparse matrices. For instance Deerwester et al. (1990) suggest, under the name of *Latent Semantic Indexing*, using a method very similar to discriminant analysis on the first principal axes of a correspondence analysis. In fact these authors use singular value decomposition, which is the basis of both correspondence analysis and principal components analysis (see Appendix 1).

Other authors emphasize the complementary between models and descriptive methods in Information Retrieval (cf. Fuhr et al., 1991) as well as the importance of visualizations that provide the best possible overviews (Fowler et al., 1991).

A case can be made for stating that the tendency is toward a gradual lessening of barriers between the various disciplines, and that "hard-nosed decisionists" are recognizing more and more the importance of the descriptive and exploratory phases of analysis.

We shall present the two areas (stylometry and global discriminant analysis) with the help of two examples of methods and applications that we feel demonstrate the differences between the techniques and the advantages of each one.

After a brief presentation of the specific statistical units and indices of stylometry (section 8.2), a first example (section 8.3) considers the stylometric problem mentioned earlier of possibly attributing a recently discovered poem to Shakespeare. A second example (section 8.5) presents a multilingual application in the context of an international survey, as an illustration of global discriminant analysis.

8.2 UNITS AND INDICES OF STYLOMETRY

Here again the basic units are derived from words (type, or graphical form), but it must be noted that many stylometric works have used counts that further divide this unit in comparing texts, authors or types.

Smith (1983) has shown the limitations of using the distribution of the number of letters per word, whereas the distribution of the number of syllables per word, systematized by Fuchs (1952), was considered useful by Brainerd (1974) for attributions to English speaking authors.

---

[1] Cf. the pioneering work of Palermo and Jenkins (1964). Cf. also Bouroche and Curvalle (1974).

Even distributions of overall frequencies of graphemes (letters and punctuation) can have an important discriminatory power among texts (Brunet, 1981: Abi Farah, 1988 [texts in Arabic]; Salem 1993). This is mostly a methodological exercise, because the specificity of the vocabularies is being perceived through the graphemes. However Brunet showed that the discriminating effect of the graphemes remains even after the most frequent words are eliminated, which makes the interpretation of those effects even more difficult.

8.2.1 Function words, speech parts

As we have previously explained, the function word cannot be formalized in a satisfactory way. Yet many authors use this concept when they rely on the common intuition that in each language there exists a list of words (function words, also called grammatical words, or tool words, or auxiliary words) that share the property of being less distinctive at the semantic level[1].

Based on such a list, Demonet et al. (1975) proposed applying a "functionality rate" to a speech or type of speech by counting the number of occurrences of words that match a list of "function words" previously compiled by them. This property has been widely applied in computerized analyses as a justification for eliminating words that are considered of little interest because of their frequent occurrence.

Conversely, the discipline that concerns itself with author attribution problems has long tended to prefer working with this type of unit, postulating that since their use was less commonplace when the text was drawn up, they could constitute an author's trademark[2]. This is the spirit of the pioneering work of Ellegard (1962) who compares the proportions of "function words" in the corpus of the "Junius Letters" (pamphlets published at the end of the eighteenth century comprising about *150 000* occurrences of words) with proportions calculated in another corpus of the same era. This operation also constitutes an important phase in the work of Mosteller and Wallace (1964) as well as that of Burrows (1992a, 1992b), Holmes (1992). More recently, Baayen et al. (1996) have underlined the importance of the syntactic environment in which a function word appears.

Benzécri carried out analyses of texts in ancient Greek (1991a), Latin (1991b) and Spanish (1992b) based on sets of function words, demonstrating the problems associated with selecting these statistical units as well as the discriminatory power of profiles of function words when they are used as active elements in a correspondence analysis.

---

1 This trait does not exclude the possibility of concentrating on such units in some areas. Think, for example, of the important role played by the tool words *for* and *against* in the analysis of political texts.

2 Cf. for example the work of Radday (1974) and Morton(1963) concerning the homogeneity of the book of Isaiah.

A categorization of the units of a text (a syntactic analysis that assigns a grammatical category to each word[1]) can be used to calculate the proportions of nouns, verbs, adjectives, etc. These proportions have also been used in stylometry. Thus, Somers (1966) asserts that the proportion of nouns indicates level of education and ease of expression. Brainerd (1974) analyzes the discriminatory power of the proportion of articles present in a text. Let us emphasize that such a categorization is necessary to identify possible function words.[2]

8.2.2. Richness of vocabulary

To conclude this brief overview of the statistical units of stylometry, we should mention richness of vocabulary indices: if V designates the number of different words, and T the total number of occurrences for each text, the first indices that were suggested were the number of distinct words V divided by the total number of occurrences T:

$$R = \frac{V}{T} \quad \textit{(type-token ratio)} \quad \text{or} \quad R' = \frac{LogV}{LogT}$$

This quotient was proposed in particular by McKinnon and Webster (1971) to discriminate among texts written under different pseudonyms by the Danish philosopher Soren Kierkegaard. When applied to a set of *16* samples that vary from *6548* occurrences to *15432*, such a ratio leads the authors to conclude that it is possible to characterize different categories of texts on the basis of this criterion alone.

In textual corpora, different parts may vary much more in length, and the two above quotients have the disadvantage of depending too much on the length of the text (briefly, for a given language, V has an upper limit, whereas T is unlimited by definition).

Simpson's (1949) index is the quotient:

$$D = \frac{\sum_r r(r-1)V_r}{T(T-1)}$$

where V_r is the number of distinct words appearing exactly r times in the text. Therefore it is the number of pairs of occurrences of the same word divided by the total number of pairs of occurrences. The numerator is no longer limited, and D depends much less on T than R or R'. D is simply the

---

1 Today such an operation cannot be totally computerized. Important progress has been made in the realm of automatic syntactic analysis of texts, as shown, for example, by the ongoing improvements in spelling correction found in most text processors.

2 Note that although isolating tool words requires categorizing and removing ambiguities in a text (as in, for example, the word *even*), some expressions contain full words that are substitutes for function words (e.g., *in fact*), that a preliminary lemmatization might obscure.

probability that two random occurrences in the corpus correspond to the same word.

This index, well known to statisticians, is very close in principle to the characteristic K introduced long before by Yule (1944) in the domain of stylistic analyses. Yule defines the two quantities S_1 and S_2 (two first moments about zero of the distribution of words), such that:

$$S_1 = \sum_r rV_r \quad (S_1 = T), \quad \text{and} \quad S_2 = \sum_r r^2 V_r \quad .$$

The characteristic K is defined as:

$$K = 10^4 \frac{(S_2 - S_1)}{S_1^2}$$

The relationship with Simpson's index D is then:

$$K = 10^4 D\left(\frac{T-1}{T}\right).$$

8.3 STATISTICAL MODELS IN STYLOMETRY: AN EXAMPLE

Until recently, most of stylometric work have made use of univariate statistical methods. Most often, some *ad hoc* discriminating parameters were calculated. However, the bulk of recent studies contains applications of multivariate methods (generally principal component analysis, factor analysis), using vectors of function words, or combinations of various stylometric indices (see Biber, 1990, 1995; Holmes, 1992), or neural network classifiers (see, for example, Tweedie *et al.*, 1996).

In addition to empirical studies that use the range of descriptive indices described previously, there are other methods that are more on the order of modelling.

An example of such methods is the previously mentioned authorship attribution analysis that was done on a poem (perhaps written by Shakespeare) discovered in 1985 (Thisted and Efron, 1987). We shall briefly cover the hypotheses, the model and the results, as well as suggesting other approaches.

8.3.1 Modelling a frequency distribution

This authorship attribution analysis is based on applications of ecological trapping models dating from the nineteen fifties. Ecologists use traps to capture specimens of various species. Let us hypothesize in an empirically reasonable and acceptable fashion that the number of specimens of a given species actually captured follows a *Poisson distribution*, whose single parameter depends upon the species. After a certain length of time, we can estimate both the number of specimens per species, and - this may appear surprising - the probable number of "trappable" species in the capturing zone. Before we present these results in greater detail for the benefit of mathematically inclined readers (the nonparametric model is presented in section 8.3.3) let us formulate them in terms of textual statistics.

The ecologist's species becomes a word (type, or graphical form), the specimen is an occurrence of this word (token). Of course, our task is to model the aggregate of the lexical distributions of a work at a given instant in time, and not throughout the process of creation. Until now we have not hypothesized independence between the numbers of occurrences per word (a hypothesis difficult to accept).

Thus this model makes it possible to estimate the evolution of the frequency distribution (i.e., new frequencies of words already used, as well as of new words) of a given *text* as the size of the text increases. Given a corpus of texts on which counts have already been made, and a new work, it is therefore possible to see whether the new work falls into the pattern already defined by all the previous works.

The model is actually valid only when the process is homogeneous with time: it does not apply *a priori* (except through empirical confirmation) when the genre changes (as with theater in verse versus theater in prose).

8.3.2 Available data for attribution problems

Starting with Spevack's (1968) work, it became possible to know the distribution of the number of distinct words as a function of their frequency of appearance in Shakespeare's work (table 8.1).

According to the counts made by this author, Shakespeare used *31 534* distinct words, with a total of *884 647* occurrences. Note that almost half of the distinct words are hapaxes.[1]

Starting with this distribution, and the model outlined above, which will later be considered in greater detail, it is possible to estimate not only the frequency of appearance of each word, but also the frequencies of new words. This led Efron and Thisted (1976) to attempt to answer the somewhat provocative question *"How many words did Shakespeare know?"*, and to estimate that the number of *additional* distinct words known by Shakespeare would have been *35 000* if the "creation" process had continued indefinitely.

1 Of course, this level of precision is misleading, because for some plays, there are entire paragraphs missing in certain editions, whereas disagreements as to the identity of words still exist for the text parts that are common to all sources.

This stretched use of the trapping model most certainly carries it beyond its range of validity, because, as we explained above, the hypothesis of homogeneity of the process is fundamental, and this is unlikely over a long period of time. In addition, assuming that all of the words to be used in the future are known at the present time constitutes an additional hypothesis. But the most surprising fact for a statistician is the rather reasonable order of magnitude of the results obtained.

Table 8.1

Frequency distribution in Shakespeare's works (partial table)

Freq.(f_r)	Count (V_r)	Freq.(f_r)	Count (V_r)
1	14376	8	519
2	4343	9	430
3	2202	10	364
4	1463	11	305
5	1043	12	259
6	837		
7	638	>12	846
Number of words		($= \Sigma_r V_r$)	31 534
Number of occurrences		($= \Sigma_r f_r V_r$)	884 647

In 1985, nine years after the above-mentioned article appeared, Gary Taylor discovered a poem with nine verses and *258* words (*429* occurrences) that might be attributable to Shakespeare. This gave an opportunity to Thisted and Efron (1987) to apply their model on this tiny potential increase in that author's work (this increase that will be denoted t in appendix 3, represents *0.05%* of the total work, in terms of occurrence).

As a precaution, the authors simultaneously analyzed seven other Elizabethan poems that are definitely attributed, of which four are Shakespeare's.

Eight Elizabethan poems

Ben Jonson	*An Elegy*
C. Marlowe	*Four poems*
J. Donne	*The Ecstasy*

Shakespeare	*Cymbeline (excerpts)*
Shakespeare	*A Midsummer Night's Dream (excerpts)*
Shakespeare	*The Phoenix and Turtle*
Shakespeare	*Sonnets (excerpts)*
Shakespeare (?)	*Taylor's Poem*

These texts were chosen because they were approximately equivalent in length and comparable in terms of genre and major thematic content.

Table 8.2 shows the distributions of the words in these eight poems compared to the Shakespearean pattern: thus Taylor's poem contains nine words never used by Shakespeare, and *140* words used by him more than *100* times.

Table 8.2

Distribution of words in the eight poems according to their frequency of appearance in Shakespeare's work

Freq.	BJon	Marl	Donn	Cymb	Mids	Phoe	Sonn	Tayl	*Total*
0	*8*	*10*	*17*	*7*	*1*	*14*	*7*	*9*	*73*
1	*2*	*8*	*5*	*4*	*4*	*5*	*8*	*7*	*43*
2	*1*	*8*	*6*	*3*	*0*	*5*	*1*	*5*	*29*
3-4	*6*	*16*	*5*	*5*	*3*	*9*	*5*	*8*	*57*
5-9	*9*	*22*	*12*	*13*	*9*	*8*	*16*	*11*	*100*
10-19	*9*	*20*	*17*	*17*	*6*	*18*	*14*	*10*	*111*
20-29	*12*	*13*	*14*	*9*	*9*	*13*	*12*	*21*	*103*
30-39	*12*	*9*	*6*	*12*	*4*	*7*	*13*	*16*	*79*
40-59	*13*	*14*	*12*	*17*	*5*	*13*	*12*	*18*	*104*
60-79	*10*	*9*	*3*	*4*	*9*	*8*	*13*	*8*	*64*
80-99	*13*	*5*	*10*	*4*	*3*	*5*	*8*	*5*	*53*
+100	*148*	*138*	*145*	*120*	*103*	*111*	*155*	*140*	*1060*
Total	*243*	*272*	*252*	*215*	*156*	*216*	*264*	*258*	*1876*

The authors calculated, for each cell of this table, the theoretical value provided by the model under the hypothesis that every one of the texts is a Shakespearean text.

Table 8.3

Theoretical and observed values for the *Taylor* poem

Frequencies	*Observed Taylor*	*Theoretical Taylor*
0	9	6.97
1	7	4.21
2	5	3.33
3-4	8	5.36
5-9	11	10.24
10-19	10	13.96
20-29	21	10.77
30-39	16	8.87
40-59	18	13.77
60-79	8	9.99
80-99	5	7.48

Thus, for example, for the *Taylor* poem, table 8.3 gives the theoretical values calculated under this hypothesis, using a formula given in appendix 3. Fitting the observed values to the theoretical values for the eight poems is carried out by using a more refined grid (the *100* values with a frequency from *0* to *99*), using a generalized linear model. The authors are thus led to reject as non-Shakespearean all of the poems that are already known to be thus, but also Shakespeare's *The Phoenix and the Turtle*.

The *Taylor* poem is not rejected. The test appears to be overly strict, since it even rejects a Shakespearean poem. But according to the authors, this finding also lends support to the belief that the new poem is part of Shakespeare's works.

More details about the nonparametric estimation model introduced above are to be found in appendix 3.

8.3.3 Other approaches to the problem

Let us consider again table 8.2 showing the frequency distributions of the words that appear in the eight poems as compared to frequency distributions of the same words in Shakespeare's works.

If we assume that the frequency distributions of the words used in Shakespeare's works are characteristic of this author, then a certain heterogeneity is to be expected in the column profiles.

Indeed, the chi-square test (the classical test of independence on contingency tables) - calculated in spite of some very small counts - is equal to, using the usual notation (see chapter 3):

$$\chi^2 = 1876 \sum_{i=1}^{n} \sum_{j=1}^{p} \left(\frac{f_{ij} - f_{i.}f_{.j}}{f_{i.}f_{.j}} \right)^2 = 104.7$$

The number of degrees of freedom of this chi-square is *77*.

Under the hypothesis of homogeneity (which corresponds here to independence of the rows and columns of the table), the probability of this value being exceeded is *0.01*. Thus this hypothesis should be rejected. One of the advantages of correspondence analysis is that it provides a visualization of how the hypothesis of independence is rejected when this occurs.

Thus figure 8.1 describes the positions of poems and frequency ranges in a principal axes plane. The controversial poem *Taylor* is near the origin, therefore near the average profile. Shakespeare's poems are lined up along a line that connects *The Phoenix* with *Midsummer*.

The Phoenix, which was rejected as non-Shakespearean by tests based on the Poisson model, is indeed close to the periphery, between *Marlowe* and *Donne,* in a zone that is abundant in new words (i.e. words whose frequency is "0" in Shakespeare) and in frequencies "2" and "10-19". Conversely, the *Midsummer* extract is abnormally devoid of new (exclusive) words.

This visualization, based on counts that have not been transformed in any way, gives rise to the idea of an enlarged empirical approach, that would consist of increasing the number of analyzed poems (Shakespeare's as well as those of other Elizabethan authors), in order to be able to determine rejection zones in a principal axes plane analogous to that of figure 8.1.

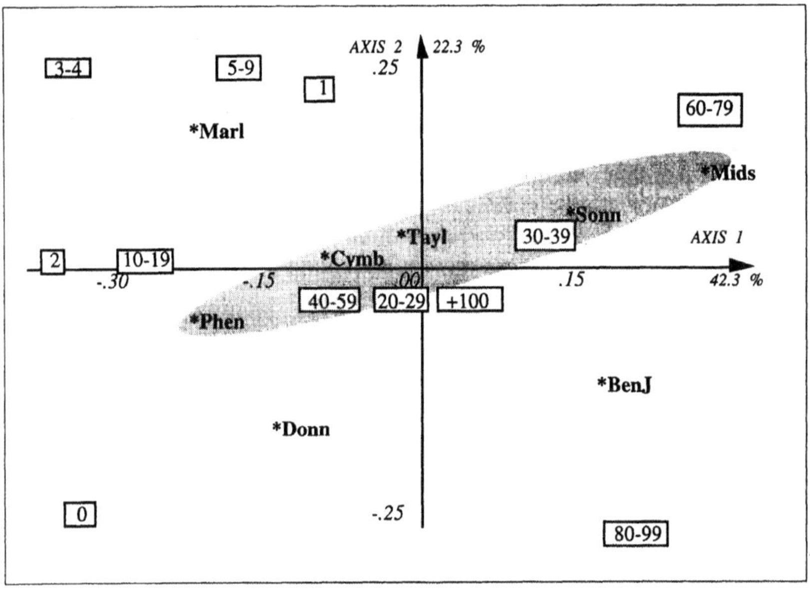

Figure 8.1

First principal axes plane of CA of (8 × 12) table: Frequency ranges × Poems (table 8.2)

The shaded area contains poems attributed to Shakespeare and Taylor poem

These zones would be based on density variations in the plane, rather than on a very small number of clues. This approach is comparable to discriminant analysis such as it is explained in the next section. However the basic variables used until now (distribution of words according to frequency ranges for a given author) are assumed to be independent of the textual content.

8.4 GLOBAL DISCRIMINANT ANALYSIS

In the analyses that follow, which are used in analyses or coding of open-ended responses in surveys or in information retrieval, the intent of discriminant analysis is to take into account both form and content, or in some cases, content alone. However the reader is reminded that taking content into account does not necessarily mean understanding it.

8.4.1 General principles

The basic principles common to the most widespread methods of discriminant analysis are:

a) There are n observations characterized by p variables, in other words, there are n points in a p-dimensional space, and those n points are classified into K groups, labeled *1* to K.

- *Example 1: 65* texts, where each is characterized by its lexical profile, calculated on the basis of the *200* most frequently occurring words in a global corpus (a space with $p = 200$ dimensions). The $n = 65$ texts correspond to $K = 2$ authors.
- *Example 2*: *1500* responses to an open-ended question characterized by *100* selected words or segments (a space with $p = 100$ dimensions). The respondents bought (or did not buy) a certain product ($K = 2$).
- *Example 3*: each observation, defined by a small number of key-words chosen among the *1500* total key-words in a thesaurus, corresponds to a document in a set that includes *10 000* documents (a space with $p = 1500$ dimensions). These documents are categorized into $K = 200$ sets of subject matter.

b) There are also m points in the same p-dimensional space as the previous n points (that is, they are characterized by the same variables or the same profiles), but these points are not classified into groups. The purpose is to assign each of these m points to the most probable group among a set of pre-defined groups.

- *Example 1* above, with $m = 12$, resembles the aforementioned problem of the *Federalist Papers*: the purpose is to determine which of two possible authors wrote each of the *12* texts.
- *Example 2*, with $m = 1$, is an attempt to predict purchase behavior on the basis of an open-ended response.
- *Example 3* is a document classification problem.

A straightforward method consists of calculating in each case the K mean profiles in the p-dimensional space and then assigning each new point to its closest profile.

The shape of each of the K *sub-configurations* of points can also be taken into account, since, as can be seen in figure 8.2, being close to the mean-point is not always the best indicator of assignment to a group.

It can be seen that the concept of predicting assignment to a group can be complex. Many technical variants exist that make it possible to interpret calculated distances in terms of probabilities.

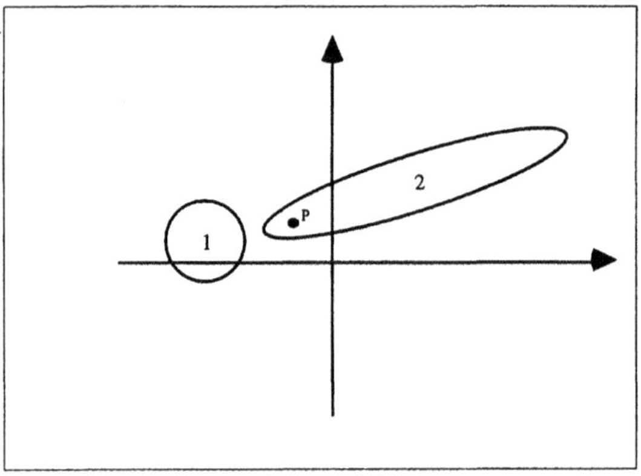

Figure 8.2
Point P is closer to center 1 than to center 2, but belongs more likely to category 2, given the observed densities

Thus in the case of figure 8.2, the *local Mahalanobis distance*, which takes into account the shape of the ellipsoids[1] corresponding to each sub-configuration, interprets point P as closer to center 2 than to center 1.

The major discriminant analysis methods that are suited to large matrices of qualitative data are:

- *discriminant analysis under conditional independence* (cf. Goldstein and Dillon, 1978).
- *discriminant analysis by direct density estimation* (Habbema et al., 1974; Aitchison and Aitken, 1976).
- *discriminant analysis by the method of nearest neighbors* (cf., for example, Hand, 1986).
- *discriminant analysis on principal coordinates* (which is used to apply to qualitative variables the same methods used for quantitative variables: factorial discriminant analysis, quadratic discriminant analysis, etc.). We will discuss this method later (Benzécri, 1977; Wold, 1976).

A thorough review of most of these methods is found in McLachlan (1992). But they can be applied only after the basic statistical units have been determined, a crucial step in textual data analysis.

---

1 The local Mahalanobis distance of point **X** to group k, which is used in *quadratic discriminant analysis*, is written $d_k(\mathbf{X}) = (\mathbf{X}-\mathbf{m}_k)' \mathbf{S}_k^{-1}(\mathbf{X}-\mathbf{m}_k)$ where $\mathbf{S}_k$ is the internal covariance matrix of group k with mean point (center of gravity) $\mathbf{m}_k$ (see: Anderson, 1984; MacLachlan, 1992).

8.4.2 Units for global discriminant analysis

The nature and the quality of the discriminant analysis can be caused to vary depending upon the choice of a basic vocabulary:

a) Words can be selected with the help of a previously established frequency threshold.
b) Words chosen can be enhanced with counts of segments or quasi-segments.
c) The text can be lemmatized (with or without elimination of function words).
d) Only the words (segments, lemmas, etc.) that characterize the groups to be discriminated can be selected beforehand.

On the basis of the working vocabulary thus created it is possible to:

e) proceed to a preliminary correspondence analysis (or to a singular value decomposition) of the table (words × individuals), and keep only the first axes (filtering and regularizing through correspondence analysis).
f) proceed to a cluster analysis of statistical units (cf. Hand, 1981; Celeux et al., 1991), and work on aggregates of units.

All of these alternatives imply a strategy to be developed by the user.

Even in the case of global discriminant analysis, varying fields of interest and areas of concern influence the choice of preliminary transformations of the basic vocabulary.

Thus, in the case of responses to open-ended questions (or information retrieval, or simply very short texts), the action of lemmatizing (*procedure c* above) or of grouping (*procedure f*), or of combining the two procedures, alters the results of calculating distances among texts. Two texts that have profiles without any common words at the outset can acquire common elements after these transformations: two variants of a lexical unit can become united into a single element.

8.4.3 Discriminant analysis and modal responses

The numbering phase and the preliminary transformation steps result in a matrix **T** of order (n,V), analogous to the matrix defined in chapter 5, whose rows are the n individuals, documents or texts and whose columns are the V different textual units (words, segments, lemmas, etc.). The K groups to discriminate define a partition of the individuals, documents or texts. They can be described by a binary matrix **Z** of order (n,K) like the one that was defined for characterizing a closed-end question with categorical variables.

We saw earlier in this chapter that the following equation can be used to calculate the aggregate lexical table **C** for any closed-end question whose responses are coded in a table **Z**:

$$C = T'Z$$

Thus lexical profiles can be compared within various groups of the population. Several tools can be used to help read the aggregated lexical tables: correspondence analysis, associated clustering techniques, lists of characteristic words and lists of modal responses. Each of these tools can serve in itself as a technique of discrimination. In particular, the process of calculating modal responses is very close to the most common forms of discriminant analysis.

Modal responses that use characteristic words

Recall that characteristic elements or words (cf. chapter 6) are words that appear with "abnormal" frequency in a part of the corpus or in the responses of a group of individuals (conversely, anti-characteristic words, which we have also called negative characteristic elements, are abnormally rare in that group).

A *modal response* for a given group has been defined as a response of an individual belonging to that group, and containing the words that characterize the group as a whole. Thus a response that happens to contain many characteristic words and very few anti-characteristic words is a very strong characteristic for that group (cf. section 6.2). Thus, to each response (or individual) i of group g can be associated the sum $s_i(g)$ of the test-values relative to the words that constitute the response (the anti-characteristic words are assigned negative test-values).

The m modal responses belong to the m individuals (m values of i) associated with the largest values of $s_i(g)$. It can be considered that $s_i(g)$ defines a *discriminant analysis rule,* since for a given individual i, the value of g for which $s_i(g)$ is maximum can define the group to which i is assigned with the highest probability.

Modal responses according to a distance criterion

Recall the selection principle presented in chapter 6. A response (or document or text) is a row-point of **T**, or a vector with V components. We learned how to calculate distances between responses and groupings of these responses, since responses (rows of **T**) and groupings of responses (columns of **C**, or rows of **C'**, transpose of **C**) are all represented by vectors in the same space. Again with the notation defined in chapter 6, the chi-squared distance between a row-point i of T and a column-point g of **C** is given by the equation:

$$d^2(i,g) = \sum_{j=1}^{p} \frac{t_{..}}{t_{.j}} \left(\frac{t_{ij}}{t_{i.}} - \frac{c_{jg}}{c_{.g}} \right)^2$$

Selecting characteristic responses translates into finding for a given group g, the individual i that minimizes the distance $d^2(i,g)$. On the other hand some usual methods of discriminant analysis consist in determining for a given individual i the value of g that minimizes the quantity $d^2(i,g)$.

8.4.4 Discriminant analysis regularized through preliminary correspondence analysis

Correspondence analyses can be performed to describe the **C** tables that are contingency tables (where the "individuals" are occurrences of words) or sparse tables **T**. They provide a visualization of the associations among words and groups or categories (see chapter 5).

They also make it possible to replace qualitative variables (simple presence or frequency of a word) with numeric variables (the values of principal coordinates), and thus to apply the classical methods of discriminant analysis (linear or quadratic) after calculating the coordinates. Correspondence analysis thus serves as a "bridge" between textual data (that are qualitative, and often sparse) and the usual methods of discriminant analysis.

But most important, as we shall illustrate in the example presented in section 8.5, a sort of filtering of information is accomplished (cf.: Wold, 1976; Benzécri, 1977) by dropping the last principal coordinates. This process strengthens the predictive power of the procedure as a whole. These properties are applied in Information Retrieval (cf. Furnas et al., 1988; Deerwester et al., 1990; Yang, 1995).

This is the meaning of the concept of regularizing discriminant analysis. Regularization techniques (cf. Friedman, 1989) strive to make discriminant analysis possible in cases that statisticians deem to be "poorly posed" (hardly more individuals than variables) or "ill posed" (fewer individuals than variables). Such cases arise when samples of individuals are small, or when the number of variables is large, which is often the situation in textual statistics. The opportunity to work on a small number of principal coordinates becomes a definite advantage.

In the examples that follow, a preliminary correspondence analysis is performed on table **T**, and further work is done on the p first principal axes. The calculation of the number of principal axes to retain can be determined so as to optimize the results of the discriminant analysis (cf. Lebart, 1992a) with the methodological precautions concerning the validity of the results that are discussed in the following section.

8.4.5 Validation of a discriminant analysis

The many options available to analysts (choosing and building statistical units, choosing a discriminant analysis technique) should not puzzle or discourage them. The quality of a discriminant analysis can be evaluated through precise criteria that are simpler than those used in exploratory methods.

In order to validate a discriminant analysis, it is necessary to perform tests on part of the sample (the test-sample) that is not used to calculate the discriminant function.

Learning sample and test sample

The discriminant function is calculated on a portion of the initial sample that is called the *learning sample (or training sample)*. One of the simplest criteria for validity is the "percent correctly classified", which is calculated for the learning sample (giving an overly optimistic view of the quality of the discriminant analysis) and for the test-sample (giving a more exact view of this quality).

This distinction among sub-samples is fundamental, and actually rather intuitive if the learning phase is described in the same terms as the learning procedures in artificial intelligence. By arriving at a maximum percentage of correctly classified cases on the learning sample, the discriminant analysis procedure uses all of the distinctive aspects of that sample, and thus includes random noise to attain its ends. In some cases, if the number of parameters is large, the percentage of correctly classified individuals can reach 100% in the learning sample, and the percentage can be very small on any other sample: this is the phenomenon of *learning by heart.* One can say that the procedure has learned without understanding, that is without capturing the general structural traits, without making a distinction between the structural information and the ancillary information in the learning sample. The true value of a discriminant analysis is only measured on one or, even better, a series of test-samples.

McLachlan (1992) performed a thorough review of the methods of error calculation in discriminant analysis. Techniques such as the Jackknife and the Bootstrap (cf. Efron (1982) for a review of these two families of methods) make it possible to estimate the degree of confidence one can have in a discriminant analysis procedure.

The same holds true for their variants, *Leaving-one-out* (or *cross-validation)* techniques[1], that consist in building *n* samples by leaving out one individual each time and applying the discriminant function to this individual (these techniques are costly and can only be applied to small samples).

In the example described in section 8.5 we apply a *leaving-x%-out* technique, which is more economical for large samples. This technique consists in drawing several test-samples where each represents a given percentage of the total sample.

This allows us to represent the quality of the prediction and the misclassification rate through *confusion matrices* that cross-tabulate the initial groups (the real ones) with the groups obtained by applying discriminant functions.

---

[1] Cf. Lachenbruch and Mickey, 1968; Stone, 1974; Geisser, 1975.

8.5 GLOBAL DISCRIMINATION AND VALIDATION

To what extent is it possible to predict that an individual belongs to a group (such as a demographic group), on the basis of his/her responses to an open-ended question? In what follows, we show through a real size example how to test the predictive power of lexical profiles constructed from responses to the open-ended questions of a survey.

In the example below, six demographic groups are obtained by cross-tabulating the categories of gender (male, female) with three age categories. These groups are common to three surveys, conducted in three countries, and the open-ended responses come in three languages.

Confusion matrices calculated on test-samples characterize the quality of the prediction as well as the structure of errors of prediction (i.e. which groups have good separation, which ones are confounded, etc.) for each of the three countries. Thus they allow us to proceed to comparisons among the three countries, in spite of the very heterogeneous (consequently not comparable *a priori*) character of the basic information.

8.5.1 Example and problem

An international survey[1] was conducted in 1989-1990 on behaviors, habits and food preferences of people in three major metropolises: Paris, New York and Tokyo.

In each of these three cities, this survey generated about one thousand in-person interviews. The questionnaire included a number of closed-end questions (common to the three countries) that described in detail the socio-demographic characteristics of the respondents, and the rhythms, timetables, locations and circumstances of meals or snacks. In addition the questionnaire had two open-ended questions, also common to the three countries.

The wording of these questions in the English version was as follows:

- *What dishes do you like and eat often?*

with a probe: *Any other dishes you like and eat often?*

- *What are some examples of what you would consider an "ideal" meal? Describe you ideal meal.*

with a probe: *Anything else that would be in your "ideal meal"?*

In each of the surveys, it was observed that the syntactic structure of the response given to this same question varied widely from one respondent to

[1] This survey was instigated by the Institute of Research on Urban Life (a Japanese research institute sponsored by Tokyo Gas Company Ltd) under the direction of H. Akuto (cf. Akuto, 1992).

the next. Some of the answers consisted of a simple list of items (names of food products or food specialties), others were in the form of relatively formal sentences, elaborating on the choices given with comments about diet or even lifestyle.

For each of the "textual corpora" collected in the three surveys, we were able to verify that the overall characteristics of the frequency distributions of the vocabulary were very similar to those that are observed in open-ended questions we had previously analyzed (some words present with very high frequency, profusion of hapaxes, etc.).

We show below some examples of responses that represent the different types of responses we have just described.

Tokyo responses (in Latin characters)

— NIMONO / EIYO NO BARANSU GA TORE, MITAME NI UTSUKUSHII KOTO.

— YASAISUPU / ESA DE NAKU SHOKUJI DE ARU KOTO, KAZOKU SOROTTE SHOKUJI O TORU, ANKA DE ARU KOTO.

— NIZAKANA, SARADA, NABERYORI, CHUKARYORI / ANZEN NA ZAIRYO O TSUKAI BARANSU NO YOI OISHIIMONO O KAZOKU YUJIN TO TANOSHIKU.

[- pot roast/balanced nutritionally, nice to look at.]

[- vegetable soup/ no food prepared sloppily, inexpensive things.]

[- cooked fish, salad, broth, Chinese food/ It is pleasant to eat with the family and with friends, good balanced meals with a lot of natural ingredients.]

Paris responses

— poulet, légumes, viande en sauce / plats surtout équilibrés avec des ingrédients le plus possible naturels.

— boeuf, canard / beaucoup de vitamines moins de matière grasses et d'additifs artificiels.

— sole au vin / repas avec entrées, plats de résistance, dessert, avec des amis dans la bonne humeur.

[- chicken, vegetables, meat with sauce / balanced meals with natural ingredients as much as possible.]

[- beef, duck / a lot of vitamins, less fat, fewer artificial additives.]

[- sole with wine sauce / meals with hors d'oeuvres, main dishes, dessert, with friends in a congenial atmosphere.]

New York responses

— baked potatoes with veal chops, fried shrimp with yellow rice, steaks, lobster with curry rice and potatoes, French onion.

— collard greens, lamb chops, rice and chicken.

— pot roast, mashed potatoes, lamb chops, salads/cucumber and pepper salad, pot roast with mashed potatoes and gravy, apple pie.

In the first analyses presented here, the responses to these two questions were grouped together and considered as one response. This grouping produces individual responses that are relatively rich at the price of a certain loss of homogeneity. Of course the three languages are analyzed separately. The Japanese responses were translated into Latin characters to be analyzable with available software.

Some categories of respondents that are common to the three surveys, such as gender and age, can be characterized in a more intrinsic fashion by their respective distances, and by the geometric configurations that summarize these distances.

Discriminant analysis makes it possible to evaluate for a given city, and thus for a given language, the predictive power of the textual response on these socio-demographic groups. It also makes it possible to answer the question: to what extent does the open-ended response of an individual give us information about his/her gender and age? This concept of predictive power can be measured in an intrinsic fashion and lead to comparisons among countries.

Here we analyze a composite variable that is based on six categories (three age categories cross-tabulated with gender). This composite variable is our "bridge" between the three surveys. The six categories are the following:

1 *less than 30 years, male*	4 *30 to 50 years, female*
2 *less than 30 years, female*	5 *over 50 years, male*
3 *30 to 50 years, male*	6 *over 50 years, female*

The idea is that starting with unknown individuals and words (at least at the outset) we expect to be able to observe and compare the configurations of known categories. This should lead us to making international comparisons, as this example attempts to demonstrate, without necessarily having to resort to a translation during early phases of the statistical treatment[1].

Tokyo sub-sample

The total text of *1008* responses contains *6219* occurrences of *832* distinct words. The analysis uses the *139* words that appear at least *7* times in the sample of *1008* responses to the two open-ended questions. These *139* words give rise to *4975* occurrences, or *80%* of the total text.

Paris sub-sample

This total text of *1000* responses contains *11108* occurrences of *1229* distinct words. The *112* words that appear at least *18* times give rise to *7906* occurrences, or *70%* of the total text.

1 Details of the analyses in the three countries are given in Akuto (1992) and Akuto and Lebart (1992).

Table 8.4

Open-ended responses of Tokyo sub-sample (latinized text) 139 most frequent words

NUM.	WORDS USED	FREQ.	NUM.	WORDS USED	FREQ.	NUM.	WORDS USED	FREQ.
1	AGEMONO	14	48	KOTO	128	95	SOBA	10
2	AJI	21	49	MENRUI	9	96	SOROTTE	24
3	ANKA	7	50	MISOSHIRU	26	97	SOZAI	20
4	ARI	15	51	MITAME	14	98	SUKI	9
5	ARU	40	52	MO	9	99	SUKINA	47
6	ATTA	11	53	MONO	209	100	SUKIYAKI	50
7	BARANSU	152	54	NABE	14	101	SUNOMONO	9
8	CHAHAN	7	55	NABEMONO	54	102	SUPAGETTEI	42
9	CHIRASHIZUSHI	7	56	NADO	7	103	SURU	24
10	CHUUKA	41	57	NAI	11	104	SUSHI	65
11	CHUUSHIN	9	58	NAKU	7	105	SUTEKI	13
12	DE	127	59	NANDEMO	15	106	TABERARERERBA	8
13	DEKIRU	8	60	NI	97	107	TABERARERU	26
14	DK	14	61	NIHON	44	108	TABERU	127
15	EIYOU	90	62	NIHONSHOKU	21	109	TABETAI	25
16	EIYOUKA	8	63	NIHONSOBA	7	110	TANOSHII	18
17	FUNIKI	10	64	NIKU	36	111	TANOSHIKU	37
18	FURAI	14	65	NIKURUI	10	112	TEMPURA	38
19	FURANSU	10	66	NIMONO	96	113	TENKABUTSU	8
20	GA	106	67	NITSUKE	9	114	TO	92
21	GOHAN	11	68	NIZAKANA	20	115	TOKI	12
22	GURATAN	20	69	NO	300	116	TOMONI	12
23	GYOUZA	15	70	O	223	117	TONKATSU	22
24	HAMBAGA	7	71	ODEN	15	118	TORE	15
25	HAMBAGU	42	72	OHITASHI	9	119	TORETA	73
26	IROIRONO	7	73	OISHII	95	120	TORETEIRU	8
27	ISSHONI	14	74	OISHIKU	75	121	TORI	8
28	ITAMEMONO	14	75	OOI	7	122	TORIIRETA	7
29	JIBUN	27	76	OOKU	16	123	TORU	23
30	JIKAN	12	77	PIZA	11	124	TSUKATTA	8
31	KAISEKI	8	78	RAISU	15	125	TSUKEMONO	14
32	KAKETE	8	79	RAMEN	27	126	UDON	15
33	KAKIFURAI	7	80	RYOURI	258	127	WA	17
34	KARAAGE	11	81	SAKANA	63	128	WASHOKU	87
35	KARADA	15	82	SAKANARUI	14	129	YAKINIKU	62
36	KARE	79	83	SAKE	13	130	YAKITORI	8
37	KARORI	11	84	SANDO	12	131	YAKIZAKANA	57
38	KATEI	14	85	SAPPARISHITA	8	132	YASAI	95
39	KATSU	7	86	SARADA	20	133	YASAIITAME	29
40	KATSUDON	7	87	SASHIMI	94	134	YASUKU	10
41	KAZOKU	64	88	SHABUSHABU	7	135	YOI	62
42	KENKOU	28	89	SHICHUU	21	136	YOKU	42
43	KENKOUTEKI	8	90	SHINSENNA	11	137	YUKKURI	17
44	KENKOUTEKINA	7	91	SHOKUHIN	8	138	YUUJIN	12
45	KI	11	92	SHOKUJI	184	139	ZAIRYOU	16
46	KISETSU	14	93	SHOKUSURU	7			
47	KONOMI	7	94	SHURUI	10			

New York sub-sample

The sub-sample of people interviewed in New York is smaller than the preceding ones (*634* persons). The corpus of open-ended responses contains *6511* occurrences of *638* distinct words. Of these *6511* occurrences, *5034* (or *77%*) concern the *83* most frequently occurring words (that appear more than *12* times). These serve as a basis for our analysis.

8.5.2 Vocabulary and analysis for Tokyo

Only the analysis pertaining to the Japanese part of the survey is presented in detail here. The analysis for New York and Paris are merely outlined. The reader can refer to the works mentioned above for detailed results of these other two surveys.

Table 8.4 displays the *139* words that appear at least *7* times in the sample of *1008* responses to the two open-ended questions.

Note the presence of function words (conjunctions, prepositions like TO, NO, NI, DE), and of foreign words adopted by the Japanese language. Some examples of word transcriptions are: BARANSU (balanced), GURATAN (au gratin), HAMBAGU, HAMBAGA (hamburger varieties), KARE (curry), SUPAGETTEI, SUTEKI, PIZA (spaghetti, steak, pizza), RAISU (rice), SARADA (salad).

Also present are some classical terms of Japanese cooking, well-known in the western world, such as SASHIMI, SUSHI, SUKIYAKI, YAKITORI, TEMPURA. Other words will be commented upon later[1].

Reading figure 8.3

Figure 8.3 represents the first principal plane resulting from the correspondence analysis of the contingency table with *139* rows and 6 columns obtained by grouping the *1008* profiles into *6* classes that correspond to the composite *age-gender* variable with six categories.

First of all, note that the lexical profiles provide a display of the relative positions of the classes where there are no cross-overs: age increases from right to left, males are at the bottom of the graph and females are at the top of the graph.

1 The latinization of Japanese writing introduced a confusion that would not have occurred with Chinese characters (Kanjis). The latinized graphical form SAKE, for example, designates *rice wine* as well as *salmon* in our coding scheme. This type of distortion of basic information only takes away a little from the richness of the aggregated lexical profiles, as the reader will be able to judge.

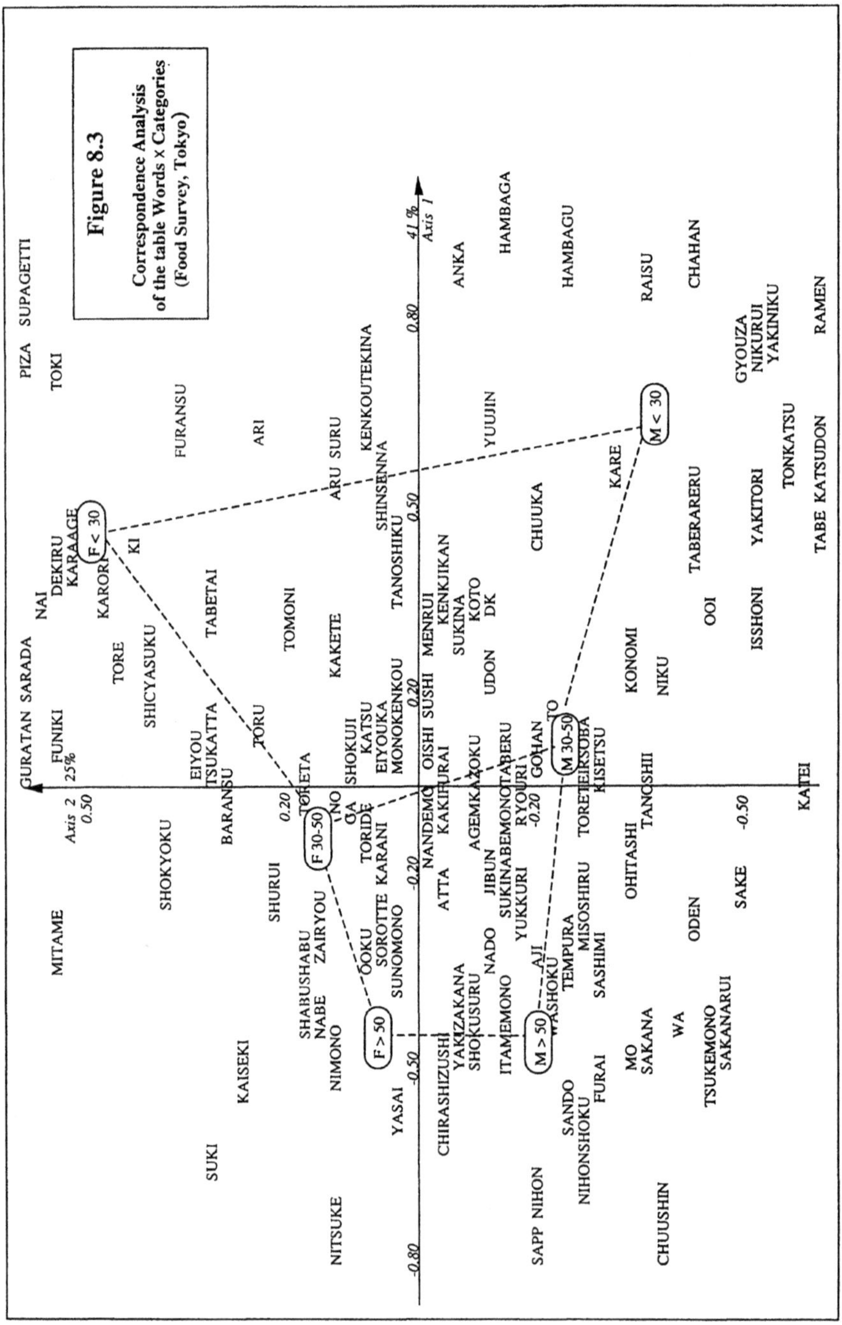

Figure 8.3

Correspondence Analysis of the table Words x Categories (Food Survey, Tokyo)

The first axis (*41%* of variance) is thus the axis that accounts for age, and the second axis (*25%* of variance) separates males from females.

In the area of young people, and mostly male youths, responses mention hamburgers, grilled meat (YAKI = grilled, NIKU = meat, YAKITORI = grilled chicken), rice, and noodles (RAMEN).

For younger and older women one can see PIZA, SUPPAGETTEI, FURANSU (French). Generally speaking, young people are mostly the ones who are interested in foreign cooking, with girls perhaps even more sensitive to European cuisine than boys.

In the area of the older classes, we find NIHON (Japanese) and the various types or preparations of fish (SAKANA or ZAKANA), with a mention of KAISEKI (refined Kyoto cuisine).

The distances represented in this map are a good representation of the distances in the initial five-dimensional space (there are six categories). In addition we observe a progressive convergence of the points relative to the two genders for a given age category as age increases, giving a trapezoidal shape to the configuration. We will revisit this observation later.

Explanations abound for this observation: traditionalism, community life leading to homogeneity of lifestyles of older people, openness of youth toward other cultures (eclectic tastes), material goods (living partly or completely outside communities, frequenting restaurants). This openness may well lead to a differentiation in culinary tastes and habits.

However the reality of the statistical phenomenon of the trapezoid still remains to be assessed. This shape may well be a distorted projection of the actual shape. First we examine the usual analyses complementary to correspondence analysis presented in chapter 6: characteristic words, which quantify in probabilistic terms the word-category associations, and modal responses, which place these words in context.

Reading table 8.5 (characteristic words)

Table 8.5 completes figure 8.3 by displaying the characteristic words of each category. To read this table, take for example the word YAKINIKU which seems to be a characteristic of males under 30 years.

On this table we read that the percentage of occurrences of this word for men under 30 (*3.95%*) is more than three times greater than its global percentage (*1.25%*). The next to last column gives this difference a test-value of *5.51*. The last column provides the associated probability threshold[1]. We observe that some words like GURATAN are characteristic of females in the first age category. On the other hand, a word such as YASAI (vegetables) describes older persons of both genders, while being much more characteristic of females.

---

1 Recall that the test-value converts a critical probability into a standardized normal variable, for easier readability: the value *1.96* corresponds to the two-tailed threshold *0.05*, whereas the value *5.51* corresponds to a probability on the order of 10^{-6}.

Table 8.5
Selection of characteristic words (Tokyo, extreme age categories)

Words	Percentage Within	Percentage Global	Frequency Within	Frequency Global	Test-value	Prob.
less than 30 years - males						
1 YAKINIKU	3.95	1.25	26.	62.	5.509	0.000
2 KARE	3.95	1.59	26.	79.	4.427	0.000
3 HAMBAGU	2.58	0.84	17.	42.	4.268	0.000
4 CHAHAN	0.91	0.14	6.	7.	3.991	0.000
5 RAMEN	1.82	0.54	12.	27.	3.812	0.000
6 RAISU	1.21	0.30	8.	15.	3.483	0.000
more than 50 years - males						
1 SASHIMI	4.00	1.89	23.	94.	3.421	0.000
2 SAKANA	2.78	1.27	16.	63.	2.938	0.002
3 CHUUSHIN	0.87	0.18	5.	9.	2.926	0.002
4 NIHON	2.09	0.88	12.	44.	2.720	0.003
5 NABEMONO	1.91	1.09	11.	54.	1.731	0.042
6 YASAI	2.96	1.91	17.	95.	1.720	0.043
less than 30 years - females						
1 SUPAGETTEI	3.57	0.84	24.	42.	6.536	0.000
2 PIZA	1.04	0.22	7.	11.	3.595	0.000
3 SARADA	1.34	0.40	9.	20.	3.239	0.001
4 GURATAN	1.04	0.40	7.	20.	2.237	0.013
5 EIYOU	2.97	1.81	20.	90.	2.160	0.015
6 TOKI	0.74	0.24	5.	12.	2.155	0.016
more than 50 years - females						
1 YASAI	4.19	1.91	38.	95.	4.910	0.000
2 NIMONO	3.64	1.93	33.	96.	3.712	0.000
3 WASHOKU	3.09	1.75	28.	87.	3.056	0.001
4 NITSUKE	0.66	0.18	6.	9.	2.905	0.002
5 NIHON	1.76	0.88	16.	44.	2.719	0.003
6 JIBUN	1.21	0.54	11.	27.	2.559	0.005

Reading table 8.6 (modal responses)

Table 8.6 shows a few characteristic responses for each of the categories being analyzed. Recall that characteristic responses for a given group are original responses selected because they are closest to the center of that group, (in the sense of the chi-square distance between the lexical profile of the response and the mean lexical profile of the group)[1].

[1] Another calculation mode consists in characterizing a response by the mean test-value of the forms which it contains. This criterion, which favors concise responses, was not used here (cf. chapter 6, section 6.2).

Table 8.6

Selection of characteristic responses (Tokyo sample, extreme age categories)

less than 30 years - males

-- 1 YAKINIKU, KARE RAISU, RAMEN, OISHII TO OMOTTE TABERARERU KOTO *[grilled meat, rice curry, noodle soup, to be able to eat tasty dishes]*

-- 2 RAMEN,YAKINIKU,KARE, OISHIKUTE RYOU NO ARU KOTO *[noodle soup, grilled meat, delicious and copious dishes]*

-- 3 KARE RAISU, HAMBAGU, SUPAGETTEI, NIKURUI, ABURAPPOI RYOURI, RAMEN NO YOUNA MONO *[rice curry, hamburger, spaghetti, meat, fat dishes, things such as soup with noodles]*

more than 50 years - males

-- 1 NABEMONO, MAGURO NO SASHIMI , YASAI TO SAKANA GA ISSHONI NATTA RYOURI *[stew, raw tuna-fish (sashimi), fish and vegetables]*

-- 2 WASHOKU, SAKANA, YASAI,TOUFU, NIMONO, JUNNIHONTEKINA MONO *[Japanese meal, fish, vegetables, tofu, family cooked food, authentic Japanese style food]*

-- 3 NABEMONO, TONKATSU, YAKIZAKANA, YASAI SAKANA CHUUSHIN NO WASHOKU *[stew, breaded pork cutlet, grilled fish, Japanese style fish and vegetables]*

less than 30 years - females

-- 1 KARE, GURATAN, PIZA, UNAGI, SUPAGETTEI, EIYOU NO BARANSU TORETA SHOKUJI *[curry, gratin, pizza, eels, spaghetti, balanced meals]*

-- 2 GURATAN, PIZA, SUPAGETTEI, CHIRASHI, KARE, HAMBAGA, SARADA, OISHIKU EIYOU GA ARI MAINICHI HENKA NI TONDA KONDATE NO SHOKUJI *[gratin, pizza, spaghetti, rice topped with various food, curry, hamburger, salad, nutritious and testy dishes, changing everyday]*

-- 3 SUPAGETTEI , BARANSU TORETA SHOKUJI *[spaghetti, balanced meals]*

more than 50 years - females

-- 1 YASAI NO NIMONO, WASHOKU *[cooked vegetables, Japanese meal]*

-- 2 YASAI NO NIMONO, SASHIMI, NIZAKANA, YAKIZAKANA, JIBUN NI ATTA SUKINA MONO O TABERU KOTO *[cooked vegetables, raw or grilled fishes, to eat one's favorite food]*

-- 3 YASAI NO NIMONO, KORUDOBIFU, SUTEKI, JIBUN DE RYOURISHITA MONO O OISHIKU AJIWAU KOTO GA DEKIREBA RISOU DE ARU *[cooked vegetables, cold beef, steaks, it is the best way to enjoy self-prepared meals]*

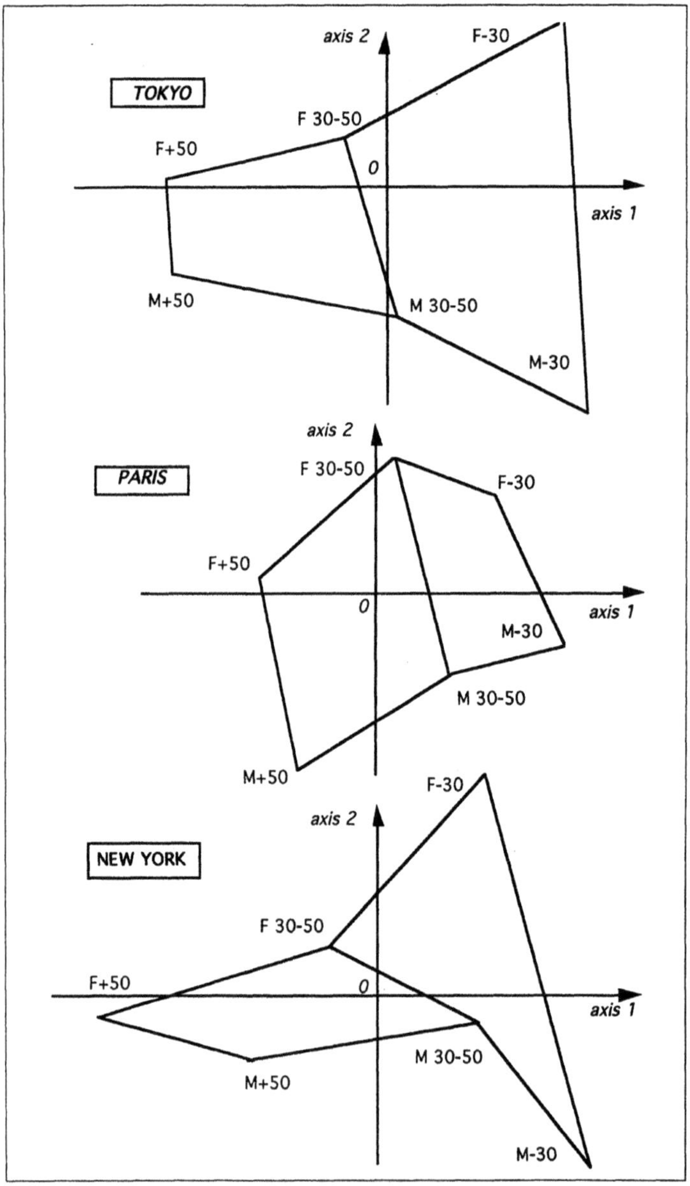

Figure 8.4
Comparison of gender-age patterns in planes (1,2) for three cities

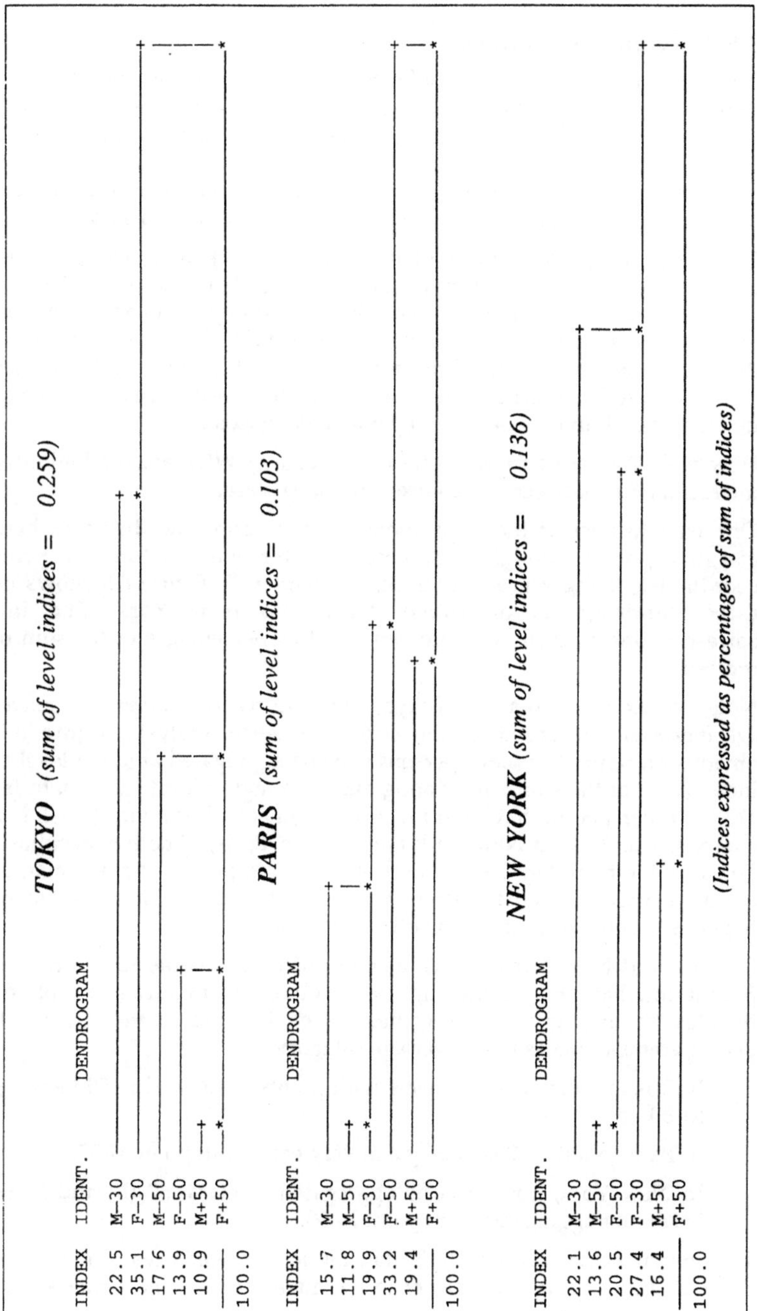

Figure 8.5: Comparison of three dendrograms (Tokyo, Paris, New York)

8.5.3 Reality of patterns

Characteristic responses are useful because they place words into their real context, and bring to the fore the complexity of the basic information. Thus, the word SUPAGETTEI, which is a strong characteristic of young females, is also sometimes found among their male counterparts. The third characteristic response among young males also contains this word. It is however found among other words that strongly characterize young males.

The analyses performed on the Paris and New York data yield patterns that are analogous to those observed in figure 8.3. For Tokyo and New York, there is a marked distance between the two youngest categories: males and females of less than *30* years, seem to give very different responses. This divergence is not seen in Paris (figure 8.4). In fact figure 8.4 sketches the three observed "patterns": note that in all three cases, ages line themselves up along axis 1 and genders are separated along axis 2.

How well do these patterns conform to reality? What are the limitations of interpretations between differences among patterns?

One way to answer these questions is to describe the distances between category points through hierarchical cluster analysis (on five principal coordinates, using Ward's criterion, cf. chapter 4). Figure 8.5 shows us the three dendrograms of Tokyo, Paris and New York. The indices corresponding to each node are expressed as percentages of the sum of the indices.

Note, for example, that for Tokyo, the *convergence between the genders as age increases* observed in the correspondence analysis (figure 8.3), is obvious on figure 8.5: older persons become aggregated at a low level of the index (*11%* of the sum), and young ones at a higher level (*22.5%*). In fact, in Tokyo, older people have the most homogeneous behaviors (level *11%*, as compared to *16%* in New York and *19%* in Paris). Another peculiarity of Tokyo is the relatively late connection of young people to the remainder of the population: the most radical cut-off is at *30* years in this city, whereas it happens around *50* years in the other two cities.

In Paris and New York, the connection of older people to the rest of the population happens at the highest level of the dendrogram. Note that females in any given age group never become connected on these dendrograms to males in a lower age category:

- In Tokyo, point F-50 connects with points M+50 and F+50 before point M-50.
- In Paris, point F-30 is aggregated very early with point M-50.
- In New York, point F-30 is aggregated to class -50 (M and F) before becoming aggregated with point M-30.

To summarize, we can conclude that the food preferences of females in all three countries bring them closer to the males of the next higher age group (from the viewpoint of response contents, characteristic words and modal responses: less interest in meats, higher interest in fish and vegetables).

The cluster analysis, based on distances in a five-dimensional space, also confirms that the relative positions of the groups (figure 8.4), and especially divergences within the younger groups in New York and Tokyo, are not artifacts due to the projection.

8.5.4 Discriminant analysis and confusion matrices

The units are chosen according to options a) and e) of section 8.4.2. The discriminant analysis is performed in the space of the words only after selection according to a frequency threshold, and after regularization by a preliminary correspondence analysis.

Regularization through principal axes methods

We shall illustrate the effect of filtering the sub-spaces resulting from the correspondence analysis by figure 8.6, which represents the success rates (that is, the percent correctly classified) of a series of discriminant analyses as a function of the number of principal axes kept (this number is on the abscissa of figure 8.6) and also as a function of the relative importance of the test samples.

This particular discriminant analysis now concerns the New York sub-sample only, the predictive variables are the *83* words, and the variable to be predicted is age divided into three categories.

- For the learning samples, (black symbols) the success rates increase continuously with the number of axes.
- For the test samples (white symbols) that measure the real discriminatory power in an actual situation, we see a saturation of the success rate at approximately *40* axes. Thus, from a certain dimension on, the additional axes do not improve the actual discriminant analysis, but they do improve the success rates of the learning sample (which are unreliable, because they are over-optimistic). This phenomenon, called *learning by heart* in Artificial Intelligence and neural networks, works within the framework of learning, but does not allow for inferences outside this framework.
- The success rate on the test sample is better if the test sample is a relatively small sample (however the difference is small between test sample representing *30%* of the total sample [square symbols] and those representing *10%* [diamond-shaped symbols]). This last point is intuitive, because learning can only improve if the learning sample is large. On the other hand if the learning sample is small, the success rates on this sample are too optimistic and therefore misleading (black squares). These three observations led us to work on a reduced number of axes (*36* in the Tokyo example) with *10%* test samples.

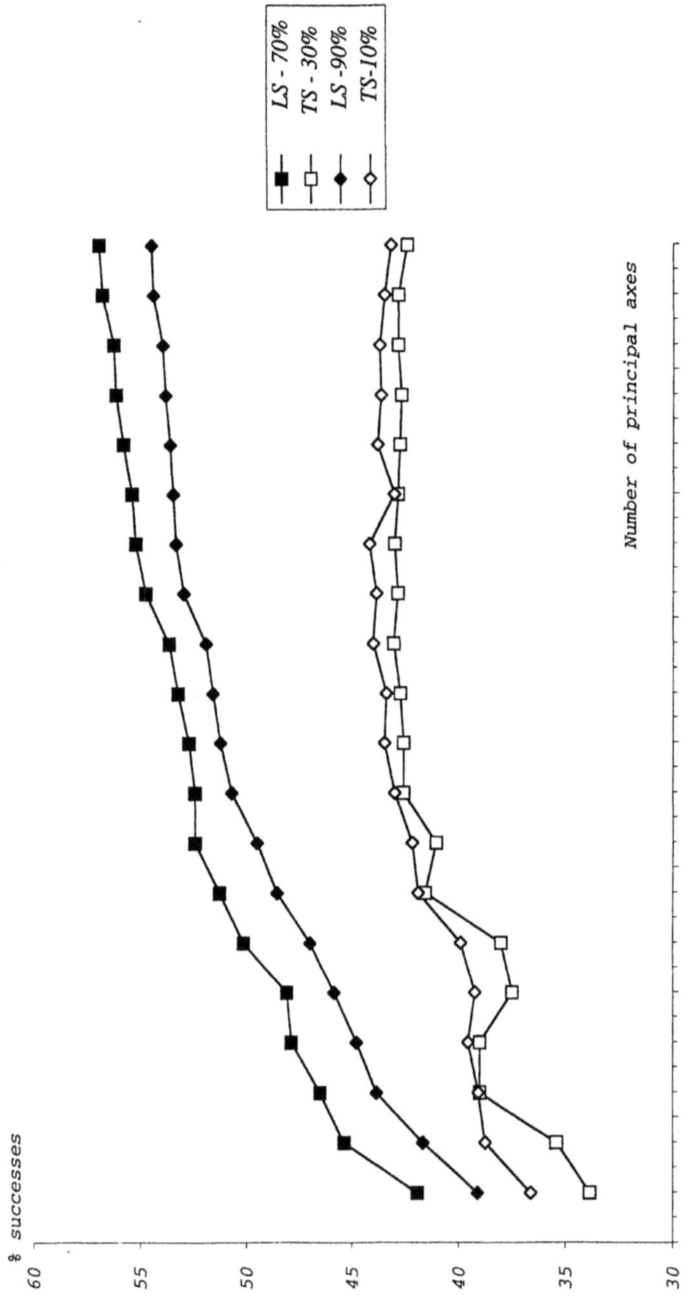

Figure 8.6 Evolution of success rate as a function of the number of axes, of size of test-sample (TS) and of learning sample (LS).

Table 8.7

Examples of confusion matrices (Tokyo)
Number of principal axes: *36*
(Test-sample: *10%* of each of *30* samples)

a - Learning sample

Actual categories [rows] and assigned categories [columns]

	M-30	M-50	M+50	F-30	F-50	F+50	Success rate
M-30	*1469*	*541*	*244*	*506*	*694*	*388*	*38.24 %*
M-50	*596*	*1457*	*620*	*676*	*1535*	*808*	*25.60 %*
M+50	*93*	*361*	*999*	*229*	*1097*	*833*	*27.66 %*
F-30	*377*	*222*	*76*	*1342*	*835*	*316*	*42.36 %*
F-50	*361*	*538*	*486*	*1146*	*2610*	*1023*	*42.34 %*
F+50	*268*	*492*	*635*	*391*	*1195*	*1841*	*38.18 %*

average success rate: *35.60* %

b - Test sample

Actual categories [rows] and assigned categories [columns]

	M-30	M-50	M+50	F-30	F-50	F+50	Success rate
M-30	*125*	*71*	*20*	*59*	*95*	*48*	*29.90 %*
M-50	*61*	*117*	*79*	*71*	*175*	*105*	*19.24 %*
M+50	*16*	*51*	*79*	*37*	*115*	*110*	*19.36 %*
F-30	*51*	*31*	*6*	*118*	*98*	*38*	*34.50 %*
F-50	*54*	*67*	*60*	*143*	*237*	*115*	*35.06 %*
F+50	*26*	*63*	*74*	*49*	*125*	*151*	*30.94 %*

average success rate: *28.13* %

Confusion matrices

Thirty test samples of approximately *100* individuals were independently drawn by random sampling without replacement from the set of *1008* individuals. For each of these *30* drawings, the six centers of gravity (mean points) of the age-gender categories are calculated in the learning sample (on *908* = *1008* - *199* individuals) in a space of *36* dimensions of the *36* first principal axes of the correspondence analysis of table T (*1008* rows, *139* columns).

Table 8.7 presents the mean confusion matrix relative to the Tokyo sample. The individuals of the learning samples (table 8.7-a) and then those of the test sample (table 8.7-b) are assigned to the nearest centers. A description of the confusion matrices relative to the three countries through correspondence analysis makes it possible to proceed to a comparison of structures that are located in different spaces (Lebart, 1992b).

Table 8.8 shows us the success rates calculated in analogous ways for the three countries. It can thus be seen that the discrimination is far superior in Tokyo than in the other cities, which was not visible on the hierarchical trees (figure 8.5). In fact, the scales of the aggregation indices are not comparable because initially neither the vocabularies used nor the samples have the same sizes.

Table 8.8

Comparison of success rates for the three cities

	Learning sample			*Test sample*		
	TOKYO	*PARIS*	*NEW YORK*	*TOKYO*	*PARIS*	*NEW YORK*
M-30	*38.24*	*20.77*	*43.45*	*29.90*	*14.44*	*33.00*
M-50	*25.60*	*13.16*	*26.23*	*19.24*	*6.89*	*20.05*
M+50	*27.66*	*31.50*	*17.08*	*19.36*	*23.13*	*10.14*
F-30	*42.36*	*29.45*	*38.00*	*34.50*	*24.59*	*30.48*
F-50	*42.34*	*29.55*	*15.78*	*35.06*	*25.28*	*12.17*
F+50	*38.18*	*30.26*	*42.00*	*30.94*	*29.41*	*37.11*
Total	*35.60*	*26.23*	*28.60*	*28.13*	*21.39*	*22.64*

Figure 8.7 gives a visualization of the row points of the two confusion matrices (active learning sample in full lines, supplementary test sample in dotted lines) relative to the Tokyo sample. The structure corresponding to the test sample is shrunk but it faithfully reproduces the structure of the learning sample.

This is a confirmation of the reality of the structure observed on figure 8.3, with this additional piece of information: the confusion appears to be very strong among men and women over 50. This finding is interpreted as follows: nothing about their responses to the open-ended question makes it possible to distinguish between these two groups, whereas the distinction between men and women was straightforward for the other two age categories.

A confusion matrix on the test-sample is a good representation of the real predictive power of the text on the age categories. Given that this representation has an intrinsic character, it can allow for comparisons among different countries and among responses expressed in different languages.

These confusion matrices can be compared to one another. This is the same as comparing the predictive powers of different texts on one categorical variable.

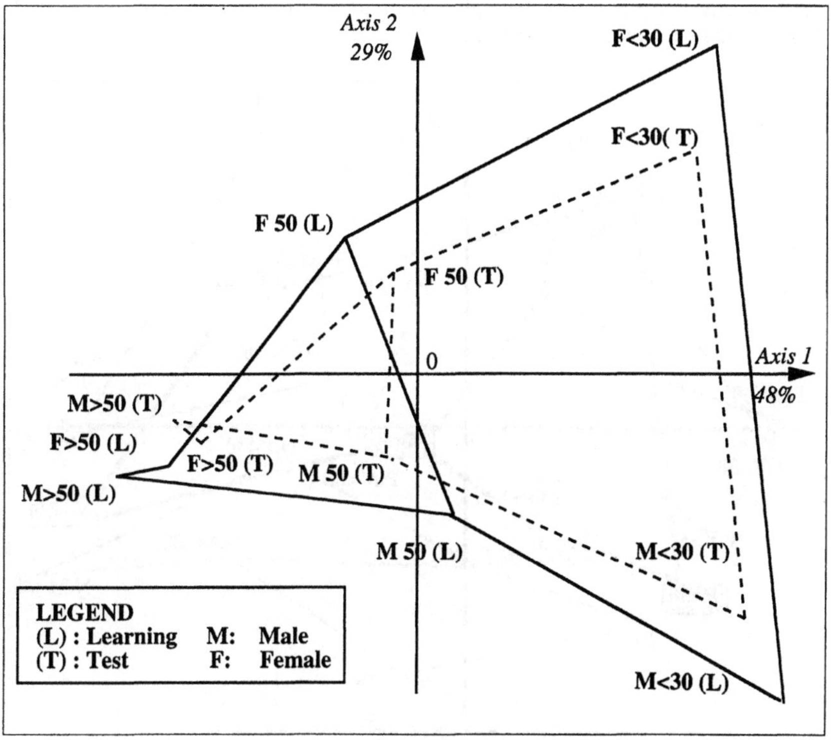

Figure 8.7
Correspondence analysis of two confusion matrices concerning Tokyo sample.

It is in this spirit that figure 8.8, which is complex only in appearance, represents a global simultaneous description of three confusion matrices (on test-samples) relative to the three cities, obtained by submitting their row-wise juxtaposition to a correspondence analysis. This stacking includes three matrices analogous to the confusion matrix of table 8.7b which only concerned Tokyo. These three matrices are built from test-samples, and therefore represent real predictive power. Thus the table analyzed includes 18 rows (6 for each city) and six columns.

Reading figure 8.8

The bold line grid connecting the framed points represents the *assigned categories* (columns of juxtaposed matrices) whereas there are three grids for the *actual categories*: New York at the bottom, Paris in dashed lines in the middle, Tokyo at the top (rows of juxtaposed matrices). Note that for each grid, young people are to the right, older people to the left, females toward the top, males toward the bottom.

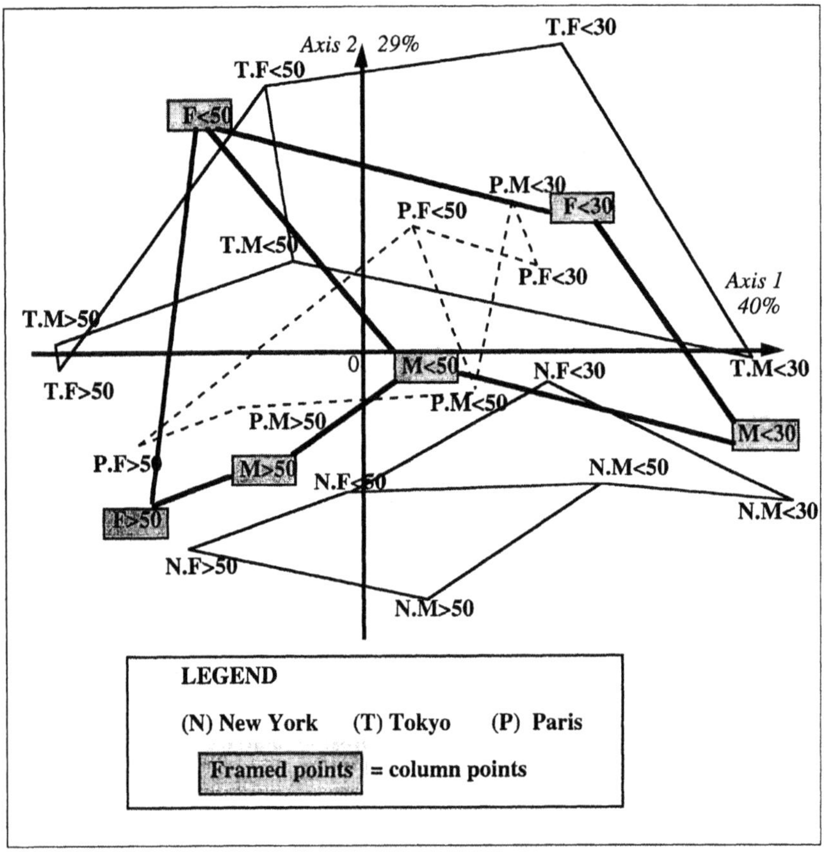

Figure 8.8
CA of juxtaposition of three confusion matrices (18 rows, 6 columns)

By looking at the position of an actual category-point relative to the bold line grid, we have an idea of how the discriminant analysis allocates the corresponding individuals. Thus, on the left, the three points F+50 corresponding to the three cities (N.F+50, P.F+50, T.F+50) are nearer to the framed point F+50 than to the other framed points, which means that the category of females over *50* years is easily recognized on the basis of the open-ended responses on food preferences in the three cities (and in three different languages). Similarly, to the right, young males in New York and Tokyo are strongly identified by their verbatim responses (points T.M-30,

N.M-30). This does not hold as true for Paris (P.M.-30) where young males are much less different from young females and from their older counterparts in terms of their responses. The dashed line grid is turned in on itself to the right side.

The first axis, linked to age, does not show discrimination among the three cities. Conversely, the second axis, linked to gender, very clearly separates Tokyo and New York. Essentially 30-50 year old females (F-50 points) and females under *30* years (F-30 points) are responsible for this state of affairs. Women in the intermediate age category in Tokyo are much better identified on the basis of their open-ended responses than are those in Paris, and even more so than those in New York. Note that this result was already visible on table 8.8 that shows the percentages of correctly classified individuals by city and by age category.

This global visualization show that the variable age remains the main source of dispersion of the data (first principal axis, before the variable country). As regards gender, its effect seems less important than that of the country. The opposition along the second axis between Tokyo and New York is stronger than the opposition between genders.

8.5.5 Conclusions to section 8.5

The possibility of comparing behaviors (in a rather straightforward manner, at least for an initial analysis) on the basis of texts in different languages obviously points to interesting new directions for research.

The phases of translation and coding of the set of responses were not necessary to proceed to descriptions and inferences that were confirmed by powerful validation procedures. But another characteristic of this example is that, in this first exploratory phase, translations would have been meaningless, since many of the basic variables that express specific food habits do not have equivalents from one country to the next.

Of course, by demonstrating the scope of possibilities of such a formal analysis of the basic information, our goal is to be better prepared for further phases of the analysis, not to circumvent them.

Appendices

Appendix 1
SINGULAR VALUE DECOMPOSITION AND CORRESPONDENCE ANALYSIS

Problem: *Assuming that the (n,p) matrix* **X** *is not a purely random matrix, is it then possible to summarize the "information" underlying* **X** *with less than np parameters ?*

A1.1 Fitting the data-points in a p-dimensional space

Let **u** be a unit vector of a p-dimensional space denoted by $\mathbb{R}^p$, ($\mathbf{u}'\mathbf{u} = 1$).
The vector **y** whose components are the *n* projections of the *n* "row-points" of **X** on the one-dimensional subspace spanned by **u** is the vector:

$$\mathbf{y} = \mathbf{X}\mathbf{u}$$

As illustrated in figure A1.1, the elements of the *i-th* row of **X** are the components of vector OM_i. The value of the *i-th* component of vector **y** is the length OH_i. Finding the best unit vector **u** according to the least squares criterion requires the minimization of $\Sigma_i\ M_iH_i^2$, or equivalently, the maximization of $\Sigma_i\ O_iH_i^2 = \mathbf{y}'\mathbf{y}$.

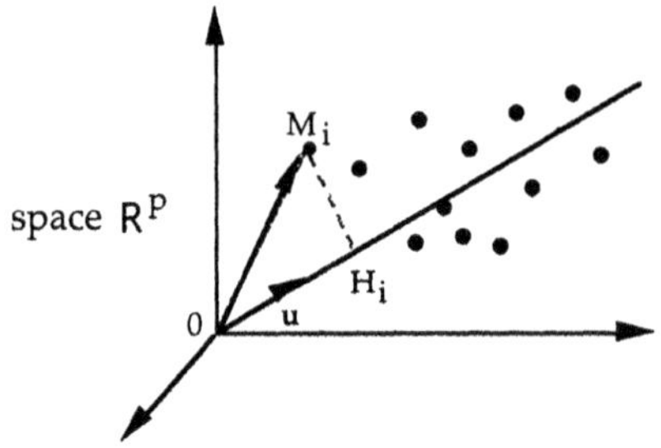

Figure A1.1
Search for the best one-dimensional subspace

The sum of squares of these projections, **y'y**, reads :

$$\mathbf{y'y} = (\mathbf{Xu})'\,\mathbf{Xu} = \mathbf{u'\,X'X\,u}$$

To determine u , we must therefore find the maximum of the quadratic form {u' X'X u} under the constraint u'u = *1*.

Let $\mathbf{u}_1$ be the solution. It can be shown that the best two-dimensional subspace is bound to contain $\mathbf{u}_1$.

More generally, if the best (q-1)-dimensional subspace is already known, the best q-dimensional subspace turns out to be spanned by $\mathbf{u}_1$, $\mathbf{u}_2$, ... $\mathbf{u}_{q-1}$, and **u** such that $\mathbf{u'u}_\alpha = 0$ ($\alpha = 1, 2, 3, \ldots q\text{-}1$), $\mathbf{u'u} = 1$, and such that {**u' X'X u**} is maximized.

This is a classical mathematical problem : $\mathbf{u}_\alpha$ is the *α-th* eigenvector of the symmetric matrix **X'X** corresponding to the eigenvalue λ_α.

Outline of proofs

- **Maximizing {u' X'X u} with u'u** = *1*

Let us compute the vector of the partial derivatives of the Lagrangian $\mathcal{L}(\mathbf{u})$ with respect to **u** :

$$\mathcal{L}(\mathbf{u}) = \mathbf{u'\,X'X\,u} - \lambda\,(\mathbf{u'u} - 1)$$

We must have : $\dfrac{\partial \mathcal{L}}{\partial \mathbf{u}} = \mathbf{0}$, which implies: $2\,\mathbf{X'X\,u} - 2\,\lambda\,\mathbf{u} = 0$.

This leads to: $\mathbf{X'X\,u} = \lambda\,\mathbf{u}$ with $\lambda = \mathbf{u'\,X'X\,u}$ (since $\mathbf{u'u} = 1$).

Therefore the maximum is necessarily an eigenvalue of the matrix **X'X** , and the absolute maximum $\mathbf{u}_1$ corresponds to its largest eigenvalue λ_1 .

- **Maximizing {u' X'X u}** with $\mathbf{u'u} = 1$ and $\mathbf{u'u}_1 = 0$.

Let us compute the derivative of the quantity:

$\mathcal{L}(\mathbf{u}) = \mathbf{u'\,X'X\,u} - \lambda\,(\mathbf{u'u} - 1) - \mu\,(\mathbf{u'u}_1 - 1)$, and let us write again : $\dfrac{\partial \mathcal{L}}{\partial \mathbf{u}} = \mathbf{0}$,

We obtain: $2\,\mathbf{X'X\,u} - 2\,\lambda\,\mathbf{u} - 2\,\mu\,\mathbf{u}_1 = 0$.

Premultiplying the two elements of this equation by $\mathbf{u'}_1$ implies that: $\mu = 0$.

This leaves: $\mathbf{X'X\,u} = \lambda\mathbf{u}$, with always: $\lambda = \mathbf{u'\,X'X\,u}$, since $\mathbf{u'u} = 1$.

λ is the second eigenvalue λ_2 of the matrix **X'X**, and $\mathbf{u} = \mathbf{u}_2$ the corresponding eigenvector.

More generally: The best q-dimensional subspace is generated by the first *q eigenvectors* of the matrix : **X'X**, corresponding to the *q* largest eigenvalues $\lambda_1, \lambda_2, \ldots\ \lambda_q$.

In the space $\mathbb{R}^p$, the vectors we considered were the n rows of the matrix **X**. If we consider now the space $\mathbb{R}^n$, generated by the p columns of the matrix **X**, shall we obtain similar results ? What relationship exists between the fits performed in these two spaces ?

A1.2 Best fit in the vector space $\mathbb{R}^n$

Let **v** be a unit vector of $\mathbb{R}^n$ ($\mathbf{v'v} = 1$). The vector **z** of the p projections of the p column-points of **X** (row-points of the transposed matrix **X'**) onto the axis **v** is the vector:

$$\mathbf{z} = \mathbf{X'v}$$

The quantity **z'z**, sum of squares of the projections, is written :

$$\mathbf{z'z} = (\mathbf{X'v})'\,\mathbf{X'v} = \mathbf{v'\,XX'\,v}$$

Maximizing $\{\mathbf{v'\,XX'\,v}\}$ with: $\mathbf{v'v} = 1$, defines again the best one-dimensional subspace, (according to the least squares criterion), in space $\mathbb{R}^n$. In analogous fashion, it can be shown that the best q-dimensional subspace is then generated by $\mathbf{v}_1$, $\mathbf{v}_2$, .. $\mathbf{v}_{q-1}$, and **v** such that: $\mathbf{v'v}_\alpha = 0$ ($\alpha = 1, 2, \ldots q\text{-}1$) and $\mathbf{v'v} = 1$, and such that the quantity $\{\mathbf{v'\,XX'\,v}\}$ **is maximum.**

Therefore $\mathbf{v}_\alpha$ is the α-*th* eigenvector of the symmetric matrix **XX'** corresponding to the eigenvalue λ'_α.

A1.3 Relationships between the fits in $\mathbb{R}^n$ and in $\mathbb{R}^p$

The relationship : $\mathbf{X\,X'\,v} = \lambda'\,\mathbf{v}$, holds in the space $\mathbb{R}^n$.

Premultiplying the two sides by the matrix **X'** , we obtain:

$$\mathbf{X'\,X\,(X'\,v)} = \lambda'\,(\mathbf{X'\,v})$$

But we have already obtained : $\mathbf{X'\,X\,u} = \lambda\,\mathbf{u}$ (with λ maximum), **u** is therefore proportional to the vector $\mathbf{X'\,v}$. Since λ is maximum: $\lambda' \leq \lambda$.

Similarly, starting from the equation $\mathbf{X'\,X\,u} = \lambda\,\mathbf{u}$, and premultiplying the two sides by **X**, we conclude that the vector **v** must also be proportional to the vector $\mathbf{X\,u}$, which leads to the new inequality : $\lambda \leq \lambda'$

In the end : $\lambda = \lambda'$.

More generally, **for every axis** α **, the non-zero eigenvalues are the same:**

$$\lambda_\alpha = \lambda'_\alpha$$

We have proved thus the proportionality of **v** and **Xu**, on the one hand, and **u** and **X' v** on the other.

$$\mathbf{v}_\alpha = k(\alpha)\;\mathbf{X\,u}_\alpha$$

$$\mathbf{u}_\alpha = k'(\alpha)\;\mathbf{X'\,v}_\alpha$$

The proportionality coefficients $k(\alpha)$ and $k'(\alpha)$ can be computed using the fact that $\mathbf{u}_\alpha$ and $\mathbf{v}_\alpha$ are unit vectors, and using also the following relationships :

$$\lambda_\alpha = \mathbf{u}_\alpha' \, \mathbf{X'X} \, \mathbf{u}_\alpha = \mathbf{v}_\alpha' \, \mathbf{XX'} \, \mathbf{v}_\alpha$$

This leads immediately to the equations : $k(\alpha) = k'(\alpha) = 1 / \sqrt{\lambda_\alpha}$, which induces the following formulas:

$$\begin{cases} \mathbf{v}_\alpha = \dfrac{1}{\sqrt{\lambda_\alpha}} \mathbf{X}\mathbf{u}_\alpha \\ \mathbf{u}_\alpha = \dfrac{1}{\sqrt{\lambda_\alpha}} \mathbf{X}'\mathbf{v}_\alpha \end{cases}$$

The original data array **X** and its transpose serve as transformation matrices linking the axes of best fit in the two vector spaces $\mathbb{R}^p$ and $\mathbb{R}^n$.

The coordinates of the projections on these axes in the two spaces $\mathbb{R}^p$ and $\mathbb{R}^n$ are respectively the rows of the vectors $\mathbf{X}\,\mathbf{u}_\alpha$ and $\mathbf{X}'\,\mathbf{v}_\alpha$.

Finally, it is easy to derive the **reconstitution formula**, or the **singular value decomposition** (SVD) (Eckart and Young, 1936) from the above formulas :

$$\mathbf{X} = \sum_{\alpha=1}^{p} \sqrt{\lambda_\alpha}\, \mathbf{v}_\alpha \mathbf{u}_\alpha'$$

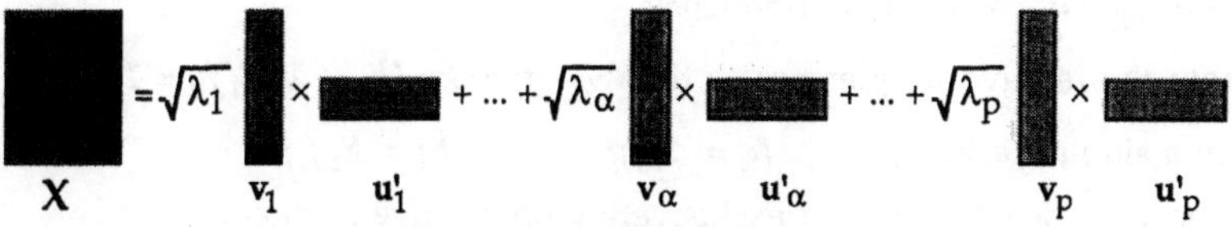

Figure A1.2
Reconstitution formula

If the smallest eigenvalues are discarded, we obtain an approximation of rank q ($q<p$) of the original matrix: $\mathbf{X} \approx \mathbf{X}^* = \sum_{\alpha=1}^{q} \sqrt{\lambda_\alpha}\, \mathbf{v}_\alpha \mathbf{u}_\alpha'$.

A1.4 Principal components analysis (PCA)

(Pearson, 1901; Hotelling, 1933)

The purely numerical process of SVD must be adapted to each particular statistical context. To describe a raw data table **R** with n rows (individuals) and p columns (variables), a preliminary transformation of the data is required.

After computing the mean: $\bar{r}_j = \frac{1}{n}\sum_{i=1}^{n} r_{ij}$

of each variable j, the table to be submitted to a SVD is the standardized table could be: $x_{ij} = \frac{r_{ij} - \bar{r}_j}{\sqrt{n}}$,

In this case, the matrix **X'X** to be diagonalized is nothing but the *covariance matrix* of the p variables.

After computing the variances: $s_j^2 = \frac{1}{n}\sum_{i=1}^{n}(r_{ij} - \bar{r}_j)^2$ of each variable j, the table to be submitted to a SVD is the standardized table :

$$x_{ij} = \frac{r_{ij} - \bar{r}_j}{s_j\sqrt{n}}$$

In this case (generally involving more heterogeneous variables) the matrix **X'X** to be diagonalized is the *correlation matrix* of the p variables.

A1.5 Correspondence analysis (CA)[1]

The data matrix **K** is a contingency table with n rows and p columns.

Notation

Let $k = \Sigma_{ij}\ k_{ij}$ be the total frequency.

We note the relative frequencies as follows : $f_{ij} = k_{ij} / k \quad (\Sigma_j \Sigma_i f_{ij} = 1)$

and, in a similar fashion, $f_{i.} = \Sigma_j f_{ij}$, $f_{.j} = \Sigma_i f_{ij}$

From now on, we deal almost exclusively with relative frequencies.

We take as the *j-th* component of the *i-th* vector of $\mathbb{R}^p$: $f_{ij} / f_{i.}$ for all $i \leq n$.

We take as the *i-th* component of the *j-th* vector of $\mathbb{R}^n$:$f_{ij}/f_{.j}$ for all $j \leq p$.

Note a fundamental difference between correspondence analysis and principal components analysis : here the transformations performed on the raw data in the two spaces are identical (because the data sets, being placed in correspondence, play similar roles).

CA can be viewed as a SVD of a transformed matrix **X** such that : $x_{ij} = \frac{f_{ij} - f_{i.}f_{.j}}{\sqrt{f_{i.}f_{.j}}}$, which appears to be equivalent to a SVD of $\hat{x}_{ij} = \frac{f_{ij}}{\sqrt{f_{i.}f_{.j}}}$.

[1] See references in main text, chapter 3.

The rationale of the method is set forth below.

Chi-square distance, distributional equivalence

The distances between two row-points i and i' on the one hand, and between two column-points j and j' on the other, are given by the equations:

$$d^2(i,i') = \sum_{j=1}^{p} \frac{1}{f_{.j}} \left(\frac{f_{ij}}{f_{i.}} - \frac{f_{i'j}}{f_{i'.}} \right)^2 \qquad d^2(j,j') = \sum_{i=1}^{n} \frac{1}{f_{i.}} \left(\frac{f_{ij}}{f_{.j}} - \frac{f_{ij'}}{f_{.j'}} \right)^2$$

Let us rephrase the property of distributional equivalence (section 3.2.2): if two rows (columns) of a contingency table are proportional (i.e., have identical profiles), then their amalgamation will not affect the column (row) distances.

Consider the case of amalgamating two rows a and b having identical profiles (we use the notation a and b instead of i_1 and i_2 to avoid sub-subscripts) into an element c whose relative frequency satisfies: $f_{c.} = f_{a.} + f_{b.}$.

In expressing the distance $d^2(j,j')$ between two columns j and j', only two terms, called T_a and T_b make use of a and b:

$$T_a + T_b = \frac{1}{f_a} \left(\frac{f_{aj}}{f_{.j}} - \frac{f_{aj'}}{f_{.j'}} \right)^2 + \frac{1}{f_b} \left(\frac{f_{bj}}{f_{.j}} - \frac{f_{bj'}}{f_{.j'}} \right)^2$$

After amalgamating rows a and b into c, the two terms are replaced by one term T_c such that:

$$T_c = \frac{1}{f_c} \left(\frac{f_{cj}}{f_{.j}} - \frac{f_{cj'}}{f_{.j'}} \right)^2$$

The property holds if: $T_a + T_b = T_c$

T_c can be written: $$T_c = f_c \left(\frac{f_{cj}}{f_c f_{.j}} - \frac{f_{cj'}}{f_c f_{.j'}} \right)^2$$

T_a and T_b can be written in a similar fashion.

Under this form, we easily obtain; $T_a + T_b = T_c$, since the profiles of a, b, and c are equal, and since: $f_{c.} = f_{a.} + f_{b.}$. Therefore, $d^2(j,j')$ remains unchanged.

This property is all the more important in the case of lexical tables, since it shows that CA is not sensitive to variations in the definition of units or categories. Units or categories with similar properties can be subdivided or amalgamated without having too much of an effect on the patterns obtained. For example, if two words a and b (e.g., corresponding to the same lemma c), have similar profiles, they can be merged without inducing a noticeable change in the CA maps.

This property is not valid for unweighted Pythagorean distances such as those used in PCA.

Goodness of fit criterion

In order to calculate goodness of fit, it is natural to give each point a mass that is proportional to its frequency.

Point i of $\mathbb{R}^p$ thus has a weight of $f_{i.}$, whereas the weight of point j of $\mathbb{R}^n$ is $f_{.j}$.

Summary

We designate by F the matrix of relative frequencies $\mathbf{F} = (1/k)\ \mathbf{K}$.

We designate by $\mathbf{D}_p$ the (p,p) diagonal matrix whose *j-th* diagonal element is $f_{.j}$; and by $\mathbf{D}_n$ the (n,n) diagonal matrix whose *i-th* element is $f_{i.}$.

With these two matrices and matrix **F**, we complete our summary of the initial elements of the analysis, in table A1.1.

Table A1.1

Coordinates, goodness of fit criterion, distances in CA.

In space $\mathbb{R}^p$	*In space* $\mathbb{R}^n$
There are n points whose coordinates correspond to the n rows of the matrix $\mathbf{D}_n^{-1}\mathbf{F}$.	There are p points whose coordinates correspond to p columns of $\mathbf{F}\mathbf{D}_p^{-1}$ (or the p rows of $\mathbf{D}_p^{-1}\mathbf{F}'$).
The goodness of fit criterion is characterized by the quadratic form whose matrix is $\mathbf{D}_n$.	The goodness of fit criterion is characterized by the quadratic form whose matrix is $\mathbf{D}_p$.
The distance is characterized by the quadratic form associated with $\mathbf{D}_p^{-1}$	The distance is characterized by the quadratic form associated with $\mathbf{D}_n^{-1}$

Fitting the data points in $\mathbb{R}^p$

We limit our presentation of to one space, $\mathbb{R}^p$, since the proofs in the other space are deduced by permutation of the indices i and j (that is, by transposition of **F**, and substitution of the matrices $\mathbf{D}_n$ and $\mathbf{D}_p$).

We wish to represent graphically the distances between profiles. Therefore we place ourselves, in both spaces, at the centers of gravity (mean points) of the configurations of points.

However, it is one of the peculiarities of correspondence analysis, one can perform the analysis either relative to the origin or to the center of gravity (provided that, in the former case, the principal axis connecting the origin to the center of gravity is ignored).

Analysis in $\mathbb{R}^p$

In this space the n points are the n rows of $\mathbf{D}_n^{-1}\mathbf{F}$. Let $\mathbf{u}$ be a unit vector for the distance in RP , that is, such that : $\mathbf{u}'\mathbf{D}_p^{-1}\mathbf{u} = 1$.

The vector of the n projections on axis $\mathbf{u}$ is written :

$$\mathbf{D}_n^{-1}\mathbf{F}\mathbf{D}_p^{-1}\mathbf{u} = \psi \qquad (1)$$

The quantity to be maximized is the weighted sum of squares :

$\psi' \mathbf{D}_n\psi = \mathbf{u}'\mathbf{D}_p^{-1}\mathbf{F}'\mathbf{D}_n^{-1}\mathbf{F}\mathbf{D}_p^{-1}\mathbf{u}$, with the constraint: $\mathbf{u}'\mathbf{D}_p^{-1}\mathbf{u} = 1$.

Therefore $\mathbf{u}$ is the eigenvector of the matrix: $\mathbf{S} = \mathbf{F}'\mathbf{D}_n^{-1}\mathbf{F}\mathbf{D}_p^{-1}$

corresponding to the largest eigenvalue λ (λ is solution of : $\mathbf{S}\,\mathbf{u} = \lambda\,\mathbf{u}$).

The general term $s_{jj'}$ of the matrix $\mathbf{S}$ to be diagonalized is written:

$$s_{jj'} = \sum_{i=1}^{n} \frac{f_{ij} f_{ij'}}{f_{i.} f_{.j}} \qquad (2)$$

This matrix $\mathbf{S}$ has the same eigenvalues as the centered matrix $\mathbf{S}^*$ (if we except the trivial eigenvalue 1 of $\mathbf{S}$ which corresponds to the eigenvalue 0 of matrix $\mathbf{S}^*$).

The general term of $\mathbf{S}^*$ is written:

$$s^*_{jj'} = \sum_{i=1}^{n} \frac{(f_{ij} - f_{i.}f_{.j})(f_{ij'} - f_{i.}f_{.j'})}{f_{i.}f_{.j}}$$

The trace of $\mathbf{S}^*$ is: $tr\, S^* = \Sigma_i\, \Sigma_j\, (f_{ij} - f_{i.}f_{.j})^2 / f_{i.}f_{.j}$

The quantity $k \times tr\, S^*$ (k multiplied by the sum of the eigenvalues) is the **classical chi-square** already mentioned (section 3.2.3).

Symmetrization of matrix S

For computational reasons, it is much more convenient to diagonalize a symmetric matrix. The previous matrix $\mathbf{S}$ is not symmetric, but the problem can easily be reduced to finding the eigenvectors and eigenvalues of a symmetric matrix $\hat{\mathbf{S}}$, whose general term is:

$$\hat{s}_{jj'} = \sum_{i=1}^{n} \frac{f_{ij} f_{ij'}}{f_{i.}\sqrt{f_{.j} f_{.j'}}} \qquad (3)$$

The two matrices $\mathbf{S}$ and $\hat{\mathbf{S}}$ have the same eigenvalues. If $\mathbf{w}$ is a unit eigenvector of $\hat{\mathbf{S}}$, then $\mathbf{u}$ such that $u_j = w_j\sqrt{f_{.j}}$ is the eigenvector of $\mathbf{S}$, with: $\mathbf{u}'\mathbf{D}_p^{-1}\mathbf{u} = 1$.

Relationship with analysis in $\mathbb{R}^n$

Thanks to the relationships between the matrices defining the profiles and distance, the symmetry between the two spaces $\mathbb{R}^p$ and $\mathbb{R}^n$ still holds.

The non-zero eigenvalues are identical in both spaces (rows and columns).

We obtain without difficulty the ***transition formulas***, using the coordinates on both principal axes corresponding to the same eigenvalue λ_α :

$$\psi_{i\alpha} = (1 / \sqrt{\lambda_\alpha})\ \Sigma_j (f_{ij} / f_{i.}) \varphi_{j\alpha} \qquad (\text{for } i = 1, \ldots, n) \qquad (4)$$

$$\varphi_{j\alpha} = (1 / \sqrt{\lambda_\alpha})\ \Sigma_i (f_{ij} / f_{.j})\ \psi_{i\alpha} \qquad (\text{for } j = 1, \ldots, p) \qquad (5)$$

Thus, if we ignore the coefficient $(1 / \sqrt{\lambda_\alpha})$, the representative points of one space are, on each axis α, the barycenters (centers of gravity) of the representative points of the other space.

We have met similar formulas previously (in the case of SVD), but here, the interpretation of such relationships is much more straightforward, since the coefficients involved are *positive, with sum equal to 1.* This is a fundamental difference with principal component analysis.

The following reconstitution formula (established by Fisher in 1940) similar to the singular value decomposition formula, is known as *Fisher's identity.*

It can be written explicitly:

$$f_{ij} = f_{i.} f_{.j}\ \Sigma_\alpha \sqrt{\lambda_\alpha}\ \psi_{i\alpha}\ \varphi_{j\alpha}$$

Taking into account the trivial solution ($\lambda_1 = 1$) presented previously, it reads:

$$f_{ij} = f_{i.} f_{.j} (1 + \Sigma_{\alpha>1} \sqrt{\lambda_\alpha}\ \psi_{i\alpha}\ \varphi_{j\alpha})$$

Coordinates

(columns "COORDINATES" of tables 3.5, 3.15, 3.17)

Diagonalization of the symmetric matrix $\hat{\mathbf{S}}$ produces a set of eigenvectors $\mathbf{w}_\alpha$, transformed into $\mathbf{u}_\alpha$. Formula (1) gives the projections of the n row-points onto the principal axes (coordinate $\psi_{i\alpha}$ of row-point i on axis α).

Explicitly: $\psi_{i\alpha} = \Sigma_j (f_{ij} / f_{i.} f_{.j}) u_{j\alpha} \quad (\text{for } i = 1, \ldots, n)$

Then, transition formula (5) gives the projections of the p column-points from $\psi_{i\alpha}$ (coordinate $\varphi_{i\alpha}$ of column-point j on axis α).

However, a simplification occurs, leading to the more direct formula:

$$\varphi_{j\alpha} = (\sqrt{\lambda_\alpha})\ u_{j\alpha} / f_{.j}$$

Distances to the origin

(column "DISTO" of tables 3.5, 3.8, 3.15, 3.17)

These squared distances are given by the formulas below, where G_p and G_n designate the centers of gravity in the spaces $\mathbb{R}^p$ and in $\mathbb{R}^n$).

Distance $d(i, G_p)$ of the row-point i in the space $\mathbb{R}^p$ is such that:

$$d^2(i, G_p) = \Sigma_j (1/f_{.j}) (f_{ij}/f_{i.} - f_{.j})^2$$

In symmetric fashion, the distance to G_n of the column-point j in $\mathbb{R}^n$:

$$d^2(j, G_n) = \Sigma_i (1/f_{i.})(f_{ij}/f_{.j} - f_{i.})^2$$

Contributions

(columns "CONTRIBUTIONS" of tables 3.5, 3.15, 3.17)

The so-called *Contributions* are measurements of the contribution of each element (row-point or column-point) to the total variance explained by each axis. Only *active elements* have non-zero contributions.

For example, the contribution of the row-point i (whose coordinates on axis α are $\varphi_{i\alpha} \sqrt{\lambda_\alpha}$) to axis α is given by:

$$c_\alpha(i) = f_{i.} \varphi_{i\alpha}^2 / \Sigma_i f_{i.} \varphi_{i\alpha}^2$$

Similarly:

$$c_\alpha(j) = f_{.j} \psi_{j\alpha}^2 / \Sigma_j f_{.j} \psi_{j\alpha}^2$$

Note that:

$$\Sigma_i f_{i.} \varphi_{i\alpha}^2 = \Sigma_j f_{.j} \psi_{j\alpha}^2 = 1$$

and that:

$$\Sigma_i c_\alpha(i) = \Sigma_j c_\alpha(j) = 1, \text{ for each axis } \alpha.$$

To make them more readable, the values of the contributions are multiplied by *100*.

Squared correlations

(column "SQUARED CORREL." of tables 3.5, 3.8, 3.15, 3.17)

The formulas are the same for both spaces $\mathbb{R}^p$ and $\mathbb{R}^n$. If we note $d_\alpha^2(i, G)$ the squared distance of point i to the center of gravity, on axis α, we have the relationship (Pythagorean theorem):

$$d^2(i, G) = \Sigma_\alpha d_\alpha^2(i, G)$$

The quantity: $\cos_\alpha^2(i) = d_\alpha^2(i, G) / d^2(i, G)$ (squared correlation)

represents the part taken by axis α in the total squared distance $d^2(i, G)$.

Note that: $\Sigma_\alpha \cos_\alpha^2(i) = 1$ (sum over all axes)

Supplementary elements

It proves useful (cf. chapter 3) to project supplementary elements (rows or columns) onto the principal axes (Cazes, 1981).

$$\psi^{+}{}_{i\alpha} = (1 / \sqrt{\lambda_\alpha}) \; \Sigma_j \, (f^{+}{}_{ij} / f^{+}{}_{i.}) \; \phi_{j\alpha}$$

$$\phi^{+}{}_{j\alpha} = (1 / \sqrt{\lambda_\alpha}) \; \Sigma_i \, (f^{+}{}_{ij} / f^{+}{}_{.j}) \; \psi_{i\alpha}$$

$\psi^{+}{}_{i\alpha}$ designates the coordinate of a supplementary row i whose profile is $(f^{+}{}_{ij} / f^{+}{}_{i.})$ on axis α, whereas $\phi^{+}{}_{j\alpha}$ designates the coordinate of a supplementary column j whose profile is $(f^{+}{}_{ij} / f^{+}{}_{.j})$ on the same axis.

Graphical display of the two sets of points on the first two axes

The superimposition of the displays of row-points and column-points is helpful in the interpretations of proximities: the proximity between two row-points can be explained by looking at the location of all the column-points. Since they involve positive coefficients, the transition relationships allow us to interpret the position of one row-point with respect to all of the column points (and vice-versa). But it is not a straightforward matter to interpret the proximity between one row-point and one column-point, since these points do not belong to the same space. This is the reason why some users prefer using separate displays.

Summary of four distinctive features of CA of contingency tables

1 - *Symmetry between rows and columns.*
There is no *a priori* distinction between variables and observations.

2 - *The chi-square distance.*
This distance is consistent with the underlying *multinomial model.* The total variance (trace) is proportional to the classical chi-square test for contingency tables.

3 - *The property of distributional equivalence.*
This property of the chi-square distance ensures a stability of results vis-à-vis amalgamations of rows and columns having similar profiles.

4 - *Equivalence between centered and non centered analyses.*
As a consequence, the *transition formulas* linking the coordinates of row-points to those of column-points only involve positive weights, allowing for a *simultaneous representation* of row-points and column-points in the same graphical display.

* *

A self-contained *Fortran* program for correspondence analysis, including diagonalization and elementary graphical displays subroutines, can be found in the book: *Multivariate Descriptive Statistical Analysis - Correspondence Analysis and Related Techniques for Large Matrices,* (Lebart *et al.*), Wiley and Sons, New York, 1984. The corresponding code is available from the authors.

Appendix 2

CLUSTERING TECHNIQUES

This appendix serves as a complement to chapter 4, with descriptions of three families of clustering algorithms :

- methods of hierarchical clustering.
- methods of clustering around moving centers, leading directly to one or several partitions of the data set.
- a mixed method, adapted to large data sets, involving the two preceding methods.

A2.1 Hierarchical clustering

The principles common to various hierarchical clustering techniques are simple.

1. At the outset we suppose that some measure of distance can be defined between pairs of elements that are to be clustered. The distances are not necessarily calculated at the outset: in this situation, it must be possible to compute or re-compute them from the coordinates of the element-points, and these coordinates must be easily accessible.

2. Then we must define *rules for calculating the distances between disjoint clusters* of elements. This between-cluster distance can generally be calculated directly from the distances of the various elements involved in the clustering.

For example, if x, y, and z are three elements, and if x and y are clustered in h, we can define the distance from this cluster h to z as the smallest distance from the elements of h to z :

$$d(h, z) = \min\{d(x, z), d(y, z)\}$$

This distance is called the *single linkage* (Sneath, 1957). Another rule that is simple and frequently used is that of the *average linkage between groups*: if x and y designate disjoint subsets containing, respectively, n_x and n_y elements, h is a subset containing $(n_x + n_y)$ elements, and we define

$$d(h,z) = \frac{n_x d(x,z) + n_y d(y,z)}{n_x + n_y}$$

Other techniques, such as *minimum variance clustering* (Ward, 1963) entail more complex formulas. They seek to optimize, at every step, the partition obtained by agglomerating two elements, using criteria that are linked to variance. These are developed below.

Clustering algorithm principles (see Hartigan, 1975; Gordon, 1981; 1987)

The general concepts we describe here are difficult to trace historically. The most systematic and earliest published explanation is perhaps that of Lance and Williams (1967). The basic algorithm for agglomerative hierarchical clusterins is as follows (Figure A2.1 describes the steps of the algorithm, starting from the set of distances displayed in table A2.1, and using the *single linkage* rule).

Table A2.1
Distances between 5 elements

	(1)	(2)	(3)	(4)	(5)
(1)	0.0	9.0	1.0	5.0	8.5
(2)	9.0	0.0	9.5	10.0	2.0
(3)	1.0	9.5	0.0	3.9	8.6
(4)	5.0	10.0	3.9	0.0	12.0
(5)	8.5	2.0	8.6	12.0	0.0

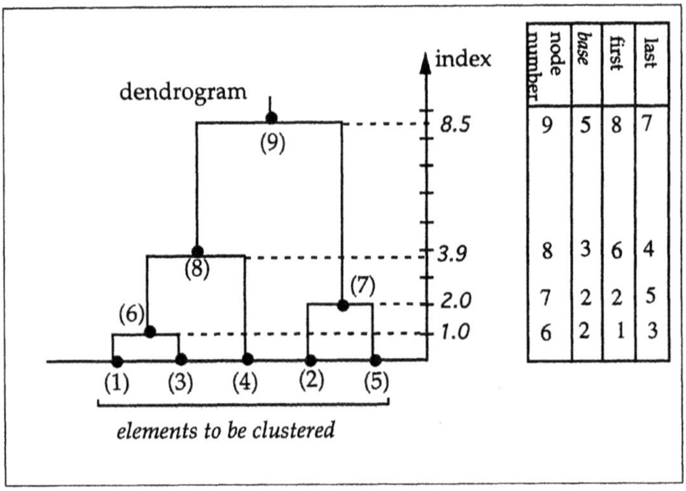

Figure A2.1
Agglomeration of 5 elements (rows or columns of table A2.1)
(single linkage method)

We call *elements* either the elements to be clustered, or the clusters of elements generated by the algorithm.

1. At step 1 there are *n* elements to cluster.

2. We find the two elements that are closest to one another and agglomerate them into a new element [elements (1) and (3) in the example of figure A1.1, such that *d(1,3) = 1.0*].

3. We calculate the distances between the new element (this new element [called (6) in the example of figure A2.1] defines a *node* of the dendrogram) and the remaining elements. We return to step 1 with only *n* -1 elements to cluster.

4. We again find the two closest elements and agglomerate them [elements (2) and (5), such that *d((2, 5) = 2.0*, creating a new element (node) number (7)]; we calculate the new distances, and repeat the process until there is only one element remaining.

Table 4.2 and 4.3 of chapter 4 are examples of clustering of rows and columns of the same contingency table. The re-computations of distances are made according to Ward's criterion. The elements have distinct weights at the outset.

Terminology

We make a few remarks to introduce the terminology associated with clustering.

The family of clusters built by such an agglomerative algorithm forms a *hierarchy*. This family has the property of containing the entire set as well as every one of the objects taken separately. The other parts of this family are then either disjoint, or included in one another (every time a new cluster is formed out of disjoint elements, the new cluster becomes itself a new element, and therefore it is necessarily included in a forthcoming cluster).

The hierarchy is an *indexed hierarchy* if a numerical value $v(h) > 0$ is associated with every part *h* of the hierarchy, such that the value is compatible with the inclusion relationship in the following way:

If *h is included into h'*, then $v(h) \leq v(h')$.

$v(h)$ is the *index* of the hierarchy.

The index $v(h)$ associated to a node *h* could be the distance between the two elements constituting *h* (figure A2.1). Such choice leads to a dendrogram that gives a good representation of the distances between elements (initial elements as well as intermediate clusters).

By "cutting" a tree or dendrogram, we obtain a *partition* in which the number of classes increases as the cut-point approaches the initial elements.

A hierarchy built form *n* initial elements allows us to obtain a set of *n* nested partitions containing 1 to *n* groups.

Single linkage clustering and minimum spanning tree

The single linkage algorithm used in our example is very simple to perform. However, it has the defect of producing "chain effects". Applying the single linkage method is practically equivalent to solving a classical problem in operations research: finding the *minimum spanning tree* on a graph (see Gower and Ross, 1969).

Minimum variance algorithms and related techniques

The minimum variance algorithm has been used for most of the examples presented in this book. It has the advantage of being compatible with correspondence analysis in the following sense: it produces indices whose sum is equal to the sum of eigenvalues (see section 4.2.1).

Here, we consider the n elements to cluster as points in a Euclidean space with p dimensions. Each point $\mathbf{x}_i$ (vector with p components) has a weight m_i. The entire mass of the set of points is noted $m = \sum_i m_i$.

The square of the distance between the points $\mathbf{x}_i$ and $\mathbf{x}_j$ is noted: $d^2(\mathbf{x}_i, \mathbf{x}_j)$

The total variance of the set of points is the quantity

$$v = \sum_{i=1}^{n} m_i d^2(\mathbf{x}_i, \mathbf{g})$$

where $\mathbf{g}$ is the center of gravity of the points.

$$\mathbf{g} = \frac{1}{m} \sum_{i=1}^{n} m_i \mathbf{x}_i$$

If there is partition of the set of objects into Q clusters, the qth cluster having a center of gravity g_q and a weight m_q, the classical decomposition of the variance v into *within-cluster* and *between-cluster* variances leads to the formula

$$v = \sum_q m_q d^2(\mathbf{g}_q, \mathbf{g}) + \sum_q \sum_{i \in q}^{n} m_i d^2(\mathbf{x}_i, \mathbf{g}_q)$$

Reduction in Variance by Agglomeration of Two Elements (Ward's Method)

Let $\mathbf{x}_i$ and $\mathbf{x}_j$ be two elements of mass m_i and m_j that are agglomerated into one element $\mathbf{x}$ of mass $m = m_i + m_j$, with

$$\mathbf{x} = (m_i \mathbf{x}_i + m_j \mathbf{x}_j) / (m_i + m_j)$$

We can decompose the variance v_{ij} of $\mathbf{x}_i$ and $\mathbf{x}_j$ with respect to $\mathbf{g}$ according to the following formula:

$$v_{ij} = m_i d^2(\mathbf{x}_i, \mathbf{x}) + m_j d^2(\mathbf{x}_j, \mathbf{x}) + m d^2(\mathbf{x}, \mathbf{g})$$

Only the last term remains if x_i and x_j are replaced by their center of gravity x. The reduction in variance Δv_{ij} is therefore

$$\Delta v_{ij} = m_i d^2(\mathbf{x}_i, \mathbf{x}) + m_j d^2(\mathbf{x}_j, \mathbf{x})$$

By replacing **x** with its value as a function of $\mathbf{x}_i$ and $\mathbf{x}_j$, we obtain

$$\Delta v_{ij} = \frac{m_i m_j}{m_i + m_j} d^2(\mathbf{x}_i, \mathbf{x}_j)$$

Thus the strategy of clustering based on the *minimum variance criterion is* as follows: instead of finding the two closest elements, we find the elements x_i and x_j corresponding to a *minimal* Δv_{ij}, which *is* the same as considering the Δv_{ij} as new dissimilarity indices.

By means of this transformation, points with small weights are more readily clustered. If we work with the coordinates of the points, we calculate the centers of gravity (x for x_i and x_j). On the other hand, if we work with the distances, it is convenient to be able to calculate the new distances from the old (as was the case in the preceding techniques).

The square of the distances between a point **z** and a cluster center **x** is written, as a function of the distances to x_i and x_j,

$$d^2(\mathbf{x}, \mathbf{z}) = \frac{1}{m_i + m_j}\left(m_i d^2(\mathbf{x}_i, \mathbf{z}) + m_j d^2(\mathbf{x}_j, \mathbf{z}) - \frac{m_i m_j}{m_i + m_j} d^2(\mathbf{x}_i, \mathbf{x}_j) \right)$$

Reciprocal neighbours algorithms

The main difficulty in building a hierarchical tree is the volume of computations. The preceding algorithms, where the nodes are constructed one by one at each step, require a number of calculations and of distance comparisons on the order of n^3 if there are n points to cluster.

More efficient algorithms greatly reduce the number of operations. The volume of calculations can decrease from n^3 to n^2, allowing the clustering of several thousand points in a reasonable amount of time. They utilize the concept of *reciprocal neighbours* (reciprocal pairs) introduced by McQuitty (1966).

The two points (or groups) *a* and *b* are reciprocal neighbours if *a* is the nearest neighbour of *b* and *b* is the nearest neighbour of *a*. At each step of the new algorithm, instead of agglomerating only the two nearest neighbours, as many new nodes are created as there are reciprocal neighbours.

At the final step, all of the points are combined into one group, and the tree is completed. The problem is then reduced to an efficient search for reciprocal neighbours (see Benzécri, 1982; *Fortran* code can be found in: Juan, 1982).

A2.2 Clustering around moving centers

This method is particularly well adapted to large numerical data sets. It is possible to read a matrix several times in sequential fashion, without requiring large amounts of computer memory. Reading directly also takes advantages of the specificity of the coding, leading to appreciable savings in computation time in the case of sparse matrices (case of binary coding for computing a *working demographic partition* (section 5.2), or case of a *direct analysis of responses* (section 5.3).

Theoretical basis of the algorithm

The algorithm underlying the programs presented here can be attributed mainly to Forgy (1965), although numerous developments (sometimes earlier, Thorndike, 1953, most often later, Ball and Hall, 1967; MacQueen, 1967; Diday, 1971) arose independently.

Let us suppose we wish to partition a set of *n elements* characterized by *p* variables. Let us also suppose that the space R^P containing the *n* elements has an appropriate *Euclidean distance* denoted by *d*. The maximum number of clusters of the partition is noted k_c.

• **STEP 0.** k_c *provisional group centers* $C^\circ{}_1, C^\circ{}_2, ..., C^\circ{}_k, ..., C^\circ{}_{k_c}$ are determined. (this could be done by a random selection, of k_c elements in the population to be clustered). These k_c centers, immediately create a partition P° into k_c clusters,

$$P^\circ{}_1, P^\circ{}_2, ..., P^\circ{}_k, ..., P^\circ{}_{k_c}$$

The assignment rule is: an element i belongs to $P^\circ{}_k$ if point i is nearer to $C^\circ{}_k$ than to all other centers. (The clusters are delimited in space by convex polyhedral divisions formed by the median planes of the segments joining all the center pairs.).

• **STEP 1.** k_c new cluster centers are determined:

$C^1{}_1, C^1{}_2, C^1{}_k, C^1{}_{k_c}$

by calculating the centers of gravity of the previous clusters $P^\circ{}_1, P^\circ{}_2, ..., P^\circ{}_{k_c}$

These new centers generate a new partition P^1 built according to the same rule as for P°. P^1 is made of the clusters : $P^1{}_1, P^1{}_2, P^1{}_k, P^1{}_{k_c}$

• **STEP m.** k_c new cluster centers are determined: $C^m{}_1, C^m{}_2, C^m{}_k, C^m{}_{k_c}$

by using the centers of gravity of the clusters: $P^{m-1}{}_1, P^{m-1}{}_2, P^{m-1}{}_k, P^{m-1}{}_{k_c}$

The algorithm stops either when two successive iterations lead to the same partition, or when a conveniently chosen criterion (for example, the measure of the within-groups variance) stops decreasing significantly, or when a previously established maximum number of iterations is reached.

In every case, the resulting partition depends upon the initial choice of the centers at step 0.

Rationale for the Algorithm

It can be shown that the within-group variance can only decrease (or remain constant) between step m and step $m + 1$. Programming conventions make it possible to force this decrease to be a strict one, and therefore to conclude that the algorithm does converge since the initial set of elements is finite.

Related Techniques

There exist many algorithms that have general principles similar to the algorithm of clustering around moving centers, but that differ in several respects.

(a) The *k-means* method introduced by MacQueen (1967) starts with a pseudo-random selection of point-centers. However the rule for calculating the new centers is not the same. Each reassignment of elements leads to a (slight) modification of the position of the corresponding center. This procedure can give a high quality partition in a single iteration. But it depends on the order of the elements on the data file, which was not the case for the technique discussed above.

(b) Other methods differ in the initial choice of the centers (equidistant elements for Thorndike, 1953), in the introduction of *thresholds,* or in the introduction of *constraints* (number of clusters, number of elements per clusters).

A2.3 Mixed clustering for large data sets

The idea of combining the approaches of Sections A2.1 (hierarchical clustering) and A2.2 (finding partitions) is not a new one: it is found, for example, under the name of *hybrid clustering* in: Wong (1982). When a large number of elements are to be clustered, the following clustering strategy is suggested. The first step is to obtain, at a low cost, a partition of the n elements into k homogeneous groups, where k is far greater than the *real* or suspected number of groups. The second step is a hierarchical clustering, where the initial elements of the dendrogram are the k groups of the partition. The third step is to visually inspect the dendrogram and cut it at the most appropriate level. The final step consists in a "consolidation" of the partition issued from the previous step, using a reassignment rule similar to that used in the moving center algorithms. The final partition of the n elements into k_f groups ($k_f < k < n$) is thus obtained.

Preliminary partition

The basic partition is calculated by using the algorithm of moving centers. The calculations are generally performed on the coordinates of the elements on the first significant principal axes of a principal axes analysis.

Hierarchical clustering of the provisional groupings

The goal of the hierarchical agglomeration phase is to further agglomerate the clusters of the basic partition, and to estimate the number of clusters. The tree is built according to Ward's criterion, which has the advantage of accounting for the size of the elements to cluster (as a weight in the calculations of the loss of variance through agglomeration).

Choosing the number of clusters

The number of clusters in the final partition of the population is defined by cutting the dendrogram. Choosing the level of the cut, and thus the number of clusters, can be done by looking at the dendrogram : the cut has to be made above the low agglomerations (low values of index), which bring together the elements that are very close to one another, and under the high agglomerations (high values of index), which lump together all of the various groups in the population. In practice, the situation is often less clearly defined (except when real, well defined groups exist in the population). The user may vacillate between two or three possible cutting levels. To date, choosing an optimum number of groups can be considered an unresolved problem (see: Gower and Banfield, 1975; Wong, 1985; Bock, 1994; Hardy, 1994).

Appendix 3

MORE DETAILS ABOUT THE NONPARAMETRIC ESTIMATION MODEL

This appendix lays out in detail the model introduced in section 8.3 of chapter 8, for the benefit of statisticians (see: Thisted and Efron, 1987). The basic hypothesis concerns the distribution of word s, a Poisson distribution with parameter λ_S. There are potentially S words, but not all of them are observed.

We designate as $G(\lambda)$ the (unknown) empirical distribution function of the numbers λ_1, λ_2, ... λ_S. Let us also designate as n_x the number of distinct words observed exactly x times.

The expected value η_x of the exact number of distinct words observed exactly x times in the past, conventionally represented here by the interval $[-1, 0]$ is written:

$$\eta_x = E(n_x) = S\int_0^\infty e^{-\lambda} \frac{\lambda^x}{x!} dG(\lambda) \tag{1}$$

Note that the expected value E_t of the number of distinct words observed at instant t but not observed beforehand (new words) is written in a simple manner.

$$E_t = S\int_0^\infty e^{-\lambda}(1 - e^{-\lambda t})dG(\lambda) \tag{2}$$

By substituting the development:

$$1 - e^{-\lambda t} = \lambda t - \frac{\lambda^2 t^2}{2!} + \frac{\lambda^3 t^3}{3!} - \ldots \tag{3}$$

in equation (2), and taking into account (1), we find that E_t can be written as the sum of the series:

$$E_t = \eta_1 t - \eta_2 t^2 + \eta_3 t^3 - \ldots \tag{4}$$

This astonishing relationship, discovered by Good and Toulmin (1956) thus gives the expected value of the number of new words for a supplementary proportion t of text, as a function of the expected values of the numbers of words η_x observed x times in the past, without the need for any hypothesis on the shape of $G(\lambda)$. The results hold regardless of this shape.

This equation was used to calculate the first term of table 8.3, substituting the theoretical values η_x with the calculated values n_x. The value of t, the proportion of new occurrences, is:

$$t = 4.85\ 10^{-4}\ (= 429 / 884\,647).$$

With such a small value, only the first term: $14\,376 \times t$ (= 6.97) is significant.

Thus with this model, the number of new words, in the case of a very small relative increase in the volume of written works, is obtained by simple proportion from the number of hapaxes previously observed.

Now we turn to the more general expected value v_x of the exact number of distinct words observed exactly x times in the interval $[-1, 0]$, and present in the interval $[0, t]$. It is:

$$v_x = S\int_0^{\infty} e^{-\lambda} \frac{\lambda^x}{x!}(1 - e^{-\lambda t})dG(\lambda)$$

The same substitution procedure using the development of (3) gives us this time a slightly more complex equation:

$$v_x = \sum_{k=1}^{\infty} (-1)^{k+1} \binom{x+k}{k} t^k \eta_{x+k}$$

To obtain an estimation of this last quantity, we can substitute the theoretical values η_x with the calculated values n_x:

$$v_x = \sum_{k=1}^{\infty} (-1)^{k+1} \binom{x+k}{k} t^k n_{x+k}$$

This is a *bayesian empirical estimator* of v_x.
For $x = 0$, (new words), we return to equation 4.

This formula is used to calculate theoretical frequencies such as those shown in table 8.3.

Appendix 4

SEARCH FOR REPEATED SEGMENTS IN A CORPUS

This appendix outlines the process of selecting unconstrained repeated segments in a corpus (cf. section 2.5.1).

In this appendix, capital letters are used to represent words.

Discarding constrained segments

The elimination of repeated segments that are constrained in the corpus considerably reduces the volume of segments while setting aside redundant results. For example, if all occurrences of segment A B C, repeated f times, are preceded by a single form X, segment A B C can be disregarded, and segment X A B C, which has exactly the same number of occurrences, can be kept. In a similar fashion, segment D E F can be set aside if its occurrences are always followed by the same form Y; in this case, only segment D E F Y is kept.

Storage of the numeric coding of the text

The numerically coded text is stored entirely in active memory at the onset of the analysis. It is represented by the array *tnum(i),* where each value is the result of coding one of the items that constitute the text.

Thus, each of the *tnum(i)* numbers corresponds to either:

- an occurrence,
- a punctuation mark,
- a symbol for identifying paragraphs, chapters, etc.

Frequency thresholds

In order to accelerate the process of searching for segments, the search is sometimes limited to repeated segments whose frequency is greater than or equal to a given threshold $fmin$ (if $fmin$ is equal to 2, the entire set of repeated segments of the corpus is obtained). The threshold also has the advantage of significantly reducing the volume of results.

Search for segments beginning with a given word A

The search for the entire set of repeated segments of a corpus can be reduced to a series of simpler sub-tasks: that of finding unconstrained segments beginning with a given form A. This same process is then repeated for all of the forms of the corpus whose frequency is greater than or equal to $fmin$.

For each form A of the text, a pivotal form of the analysis, whose frequency is greater than f_{min}, we begin by storing the addresses of each of the occurrences of form A in an array *list(j)*.

- *list(j)* contains the address of the jth occurrence of A.

These addresses are then sorted according to the alphabetical order of the forms that follow the pivotal form.

If the form following A is unchanging, the addresses are then sorted according to the second following form, etc.

The order thus obtained is the order in which the context rows appear when a concordance is carried out that is sorted on the context that follows a pivotal form (cf., for example, section 2.4.2, table 2.6).

- *list(j)* now makes it possible to access the occurrences of pivotal form A in the order of a concordance sorted on the form that follows A.

After the sort, the occurrences of form A that correspond to a single repeated segment (starting with A) are placed one after another, as shown in table A4.1.

Table A4.1
Searching for repeated segments beginning with form A
(first seven series beginning with A*)*

S1	A B
S2	A B
S3	A B C
S4	A B C D E F G
S5	A B C D E F G
S6	A B C D E D H
S7	A C B A

In practice, a series is interrupted every time a form is encountered whose frequency is less than f_{min}, because such a form cannot be included in a segment of frequency at least equal to f_{min}.

A series is also interrupted when a punctuation mark is encountered.

In order to find segments of length 2, 3, 4, etc., and their respective frequencies, it is sufficient to compare series of forms thus represented in this order.

It is not necessary to store table A4.1 in memory.

It can be accessed dynamically through arrays *list(j)* and *tnum(i)*.

Two successive series Sa and Sb can be compared to identify the first position x (let us name it *accident*) for which an occurrence of the second segment does not correspond to the occurrence of the preceding segment.

In the comparison of the two series S5 and S6 of the table above, for example, *accident* is located in position number 6 (D ≠ F).

	1 2 3 4 5 6 7
S5	A B C D E F G
S6	A B C D E D H
	≠
	x=6

An *accident* of this type in position x indicates that the search has ended for repeated segments that begin with A B C D E F and whose length is greater than or equal to x.

In our example, the process of searching for segments whose length is greater than or equal to x is then re-initialized with the values for segment A B C D E D H.

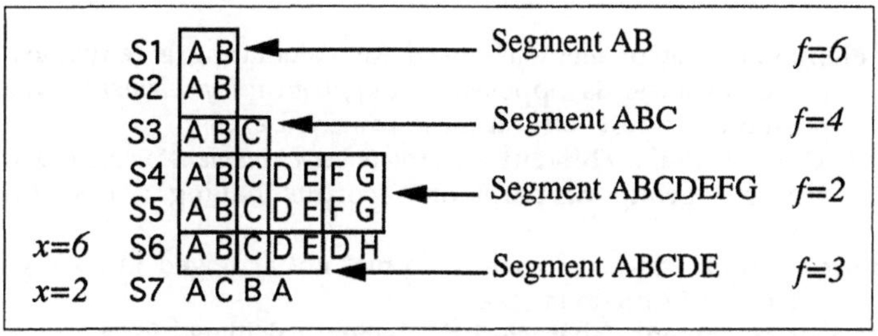

Figure A4.1
Searching for repeated segments
(f designates the frequency of a segment)

For each repeated segment thus inventoried, we then check whether each of its occurrences is either preceded or followed by the same form, and whether its frequency is at least equal to f_{min}. This is done by referring to the occurrences of the numerically coded text that are stored in memory.

If these conditions are met, we have found an unconstrained segment of frequency greater than or equal to f_{min}, that starts with form A.

This same process is repeated for all the forms of frequency greater than or equal to f_{min}.

* *

Using these principles, briefly outlined here, it is possible to extract the repeated segments from a corpus having several hundred thousand occurrences on an ordinary personal computer, in a few minutes.

Academic software is available from the authors.

Glossary

Some of the terms of this glossary are used within very different disciplines (statistics, lexicometrics, stylometry, information retrieval, quantitative linguistics, socio-economic surveys) that have their own terminologies. Here, we attempt to specify the meanings with which terms are used throughout this book.

NB: Asterisks refer the reader to another entry in this glossary.

active elements - set of elements used for calculating eigenvalues* and principal axes, as opposed to *supplementary elements** that are positioned *a posteriori* on these principal axes.

adjacent characteristic element - *characteristic element** concerning an adjacent group of parts of a corpus having a *longitudinal partition**.

alphabetical index - index* in which forms* are arranged in *lexicographic order** (as in dictionaries).

characteristic increment - characteristic element calculated for a corpus part relative to all preceding parts in a chronological corpus.

characteristic element (word, lemma, segment) -(of a part) textual unit that is unusually frequent (positive characteristic unit) or unusually rare (negative characteristic unit) in a specified part of a corpus.

chi-square distance - distance between profiles* of frequencies, used in *correspondence analysis** and in some *cluster analysis** algorithms. This distance has the property of *distributional equivalence**.

closed question - (or: *pre-coded question*) in sample survey questionnaires, question provided with a set of fixed alternatives from which respondents can choose.

cluster analysis - statistical technique for grouping elements in such a way that elements in the same cluster are similar to one another, and dissimilar to elements in other clusters.

co-occurrence - simultaneous, but not necessarily contiguous, presence of occurrences of two given words in a fragment of text (sequence, sentence, paragraph, neighbourhood of occurrence, corpus part, etc.).

concordance - set of rows of contexts around a given pivotal-word.

contingency table - (syn.: for table of frequencies or cross-tabulation) table whose rows and columns represent respectively the categories of two categorical (nominal) variables, and whose general term represents the number of individuals associated with each pair of categories.

contribution - contribution (of an *active element** i to a principal axis α) measures the importance of element i in the construction of axis α. For a given principal axis, the sum of the contributions of the elements (over all the active elements) is equal to *100*.

corpus - limited set of texts upon which the study of a linguistic phenomenon is based. In lexicometrics, a set of texts that are combined for comparison purposes, serving as a basis for a quantitative study.

correspondence analysis - (CA) principal axes method applied to *contingency tables**. Correspondence analysis is mainly characterized by: 1) a symmetric processing of rows and columns; 2) the use of a special distance called *chi-square distance**; 3) simple *transition formulas** allowing for a simultaneous representation of rows and columns.

delimiting/non-delimiting characters - differentiation established among all characters within a text in order to make it possible for computerized procedures to segment the text into occurrences*.

There are several types of delimiting characters:

- **occurrence delimiters**, which are generally: blanks, common punctuation marks.

- **sequence delimiters**: subset of occurrence delimiters consisting of weak and strong punctuation marks.

- **sentence separators**: subset of sequence delimiters consisting of strong punctuation marks.

dendrogram - (syn.: hierarchical tree) graphic representation of a *hierarchical cluster analysis**, showing the progressive inclusion of clusters.

discourse/language - language is an abstract object that can only be understood in its oral or its written implementation; "discourse" is a convenient term that encompasses both implementations.

distribution - (of occurrences* of an element in the parts of a corpus*) the series of p numbers (p= number of parts in the corpus) constituted by the sequence of sub-frequencies* of that element in each of the parts, shown in the same order as the parts.

distributional equivalence - property of the *chi-square distance**, and thus of the representations obtained through *correspondence analysis**, stating that: if two rows (*resp.* columns) of a *contingency table** are proportional, then their amalgamation does not affect the column (*resp.* row) distances.

eigenvalue - quantities that make it possible to determine the relative importance of successive principal axes in a principal axes decomposition. The eigenvalue noted λ_α measures the variance of elements on axis α.

form - type corresponding to identical occurrences* in a corpus of texts.

function word - word which indicates a grammatical relationship (as opposed to content word) (syn.: tool word, grammatical word, auxiliary word). In practice, function words are often defined through a closed list of articles, conjunctions, prepositions, etc..

graphical form - syn.: form, type.

hapax - Gr. hapax (legomenon), "something stated once". Word whose frequency is equal to one in the corpus (corpus hapax) or in one of its parts (part hapax).

hierarchical cluster analysis - (syn.: *hierarchical classification*) special *cluster analysis** technique yielding, through progressive agglomeration, clusters that have the property of being either disjunct or included in one another.

hierarchical index - index* in which words are arranged in *lexicometric order**.

index - (lexicometrics) - list consisting of a rearrangement of the words of a text grouping together references* relative to the occurrences* of a given word.

index - (*cluster analysis**; *hierarchical clustering**) - The index attached to a node (also: the *height* of the node) formed by the agglomeration of two elements i and i' is a quantity that characterizes the level of the node* in the hierarchy. The smaller the index, the more the elements i and i' are regarded as being similar.

lemmatization - grouping of occurrences of text into a canonical form (generally on the basis of a dictionary). In English, this grouping is generally done in the following way:
- verbal forms in the infinitive,
- nouns in the singular,
- elided forms to a form without elision.

length - (of a corpus, of a part of that corpus, of a text fragment, of a segment, etc.) number of occurrences (token) contained in that corpus (or of that part, fragment, etc.). Syn.: size.

lexical table - (LT) *contingency table** cross-tabulating words and parts of a corpus. The general term $k(i,j)$ of the lexical table is equal to the number of times that word i is found in part j of the corpus.

lexicographic order - order in which words are arranged in a dictionary.

lexicometric order - order resulting from sorting the words of a corpus* in order of decreasing frequency; words of same frequency are arranged in *lexicographic order**.

lexicon - entire set of words of a language.

longitudinal partition - ordered partition of a corpus*.

maximum frequency - frequency of the most frequently occurring word in a corpus*.

modal response - (of a group of individuals, of a part of a corpus*) response selected as being representative of a group (or of a part) in general on the basis of the *characteristic words** it contains.

multiple correspondence analysis - *correspondence analysis** method that applies to analysing complete disjunctive tables, binary tables whose rows are individuals or observations, and columns the juxtaposition of response categories of questions (the response categories being mutually exclusive).

negative characteristic element - for given word *i* and part *j*, word *i* is said to be a negative characteristic element of part *j* (or an "anti-characteristic word" of that part) if its frequency is "unusually low" within that part.

node - grouping created at each step of a *hierarchical cluster analysis** algorithm by agglomeration of the two closest elements.

occurrence - series of non-delimiting characters whose boundaries at both ends are *occurrence delimiters** (syn.: token).

open-ended question - (syn.: open question) question asked without a pre-established response grid, leading to verbatim responses.

partition - (of a corpus of texts) division of a corpus into *parts* composed of *consecutive text fragments* that do not have a common intersection and whose union is equal to the corpus.
(of a set, of a sample) division of a set of individuals or observations into disjunct *groups* whose union is equal to the complete set.

percentages of variance - quantities proportional to eigenvalues*, whose sum is equal to *100*.

positive characteristic element - for given word *i* and part *j*, word *i* is said to be a positive characteristic element of part *j* (or a *characteristic word** of that part) if its sub-frequency is "unusually high" within that part.

post-coding - manual operation consisting in finding the main categories of verbatim responses in a sub-sample of responses, then in closing the corresponding open-ended question by assigning all of the responses to these categories.

principal axes methods - (or: *eigen-analyses*) family of multivariate statistical methods whose aim is to extract principal axes that approximately summarize the information contained in the initial data table.

profile - (of a row or of a column) vector composed of the counts contained in row (or column) of a contingency table, divided by the sum of the counts for that row (or column).

references - system of numeric coordinates used to find within the original text each of the occurrences* resulting from a segmentation (e.g., volume, page, line, position within line) or to locate this occurrence among pre-defined categories (author, year published, mention, emphasis, etc.).

relative frequency - frequency of a textual unit in a corpus* (or in one of its parts) divided by the size of the corpus (or of the part).

repeated segment - series of consecutive forms whose frequency is greater than or equal to 2 in the corpus*.

response item - (or response category) pre-determined response element in a *closed question**.

segmentation - set of operations performed through computerized procedures that result in dividing a text stored on a machine-readable medium into distinct units.

sentence separators - *see*: *delimiting characters**.

sequence - series of occurrences of a text not separated by a sequence delimiter*.

sequence delimiters - *see*: *delimiting characters**.

size - *see*: length*.

squared correlation - (syn.: relative contribution) parameter that show the importance of the different axes in explaining the variance of an element. For a given element, the sum of squared correlation over all axes is equal to *1*.

supplementary (illustrative) elements - set of elements that are positioned on principal axes *a posteriori*, but do not participate in calculations of these principal axes. A supplementary element can be considered as an active element with a null weight.

test-values - quantities that make it possible to determine the significance of the position of a *supplementary (or illustrative) element** on a principal axis. Briefly, a test value can be considered as a standardized normal variable under the hypothesis of independence between the element and the principal axis.

token - throughout this book, synonym of occurrence*.

transition formulas - formulas linking together the coordinates of row-points and column-points on a principal axis in correspondence analysis. Such formulas account for the simultaneous representation of rows and columns.

type - *see:* form*.

***type T* variables** - variables whose frequency is approximately proportional to the rate at which a text lengthens (ex: maximum frequency).

***type V* variables** - variables whose growth has a tendency to diminish as the text lengthens (ex: number of forms, number of hapax).

verbatim response - (syn.: free response) response to an open-ended question.

vocabulary - set of distinct words*(types) found in a corpus.

word - throughout this book, synonym of form* or type.

References

Abi Farah A. (1988): Reconnaissance de l'auteur d'un texte d'après les caractères utilisés, *Les Cahiers de l'Analyse des Données*, XIII, 1, 95-96.

Adams L.L.(1975): *A statistical analysis of the book of Isaiah in relation to the Isaiah problem*, Brigham Young University, Provo.

Aitchison J., Aitken C. G. G. (1976): Multivariate binary discrimination by the kernel method, *Biometrika*, 63, 413-420.

Akuto H. (Ed.) (1992): *International Comparison of Dietary Cultures*, Nihon Keizai Shimbun, Tokyo.

Akuto H., Lebart L. (1992): Le repas idéal. Analyse de réponses libres en anglais, français, japonais, *Les Cahiers de l'Analyse des Données*, XVII, 3, Dunod, Paris, 327-352.

Anderson T.W. (1984): *An Introduction to Multivariate Statistical Analysis* (Second edition), J. Wiley, New York.

Baayen R.H., Van Halteren H., Tweedie F. (1996): Ouside the cave of shadows: Using syntactic annotation to enhance authorship attribution, *Literary and Linguistic Computing*, 11, 121-131.

Babeau A., Lebart L. (1984): Les conditions de vie et les aspirations des Français, *Futuribles*, Avril, 37-51.

Ball G.H., Hall D.J. (1967): A clustering technique for summarizing multivariate data, *Behavioral Sciences*, 12, 153-155 .

Bartell B.T., Cottrell G.W., Belew R.K. (1992): Latent semantic indexing is an optimal special case of multidimensional scaling, *Proceedings of the 15th Int. ACM-SIGIR Conf. on Res. and Dev. in Information Retrieval*, Belkin N. *et al.* (Eds), 161-167, ACM Press, New York.

Baudelot C. (1988): Confiance dans l'avenir et vie réussie (Réponse à un questionnaire), *Recherches économiques : études en hommage à Edmond Malinvaud*, Economica et EHESS, Paris.

Becue M. (1988): Characteristic repeated segments and chains in textual data analysis, *COMPSTAT, 8th Symposium on Computational Statistics*, (short papers), Physica Verlag, Vienna.

Becue M., Peiro R. (1993): Les quasi-segments pour une classification automatique des réponses ouvertes, in *Actes des Secondes Journées Internationales d'analyse des données textuelles*, (Montpellier), ENST, Paris, 310-325.

Belson W.A., Duncan J.A. (1962): A Comparison of the check-list and the open response questioning system, *Applied Statistics*, 2, 120-132.

Benveniste E. (1966): *Problèmes de linguistique générale*, Gallimard, Paris.

Benzécri J.-P.(1977): Analyse discriminante et analyse factorielle, *Les Cahiers de l'Analyse des Données*, II, 4, 369-406.

Benzécri J.-P. (1973): *La taxinomie* (T. I) ; *L'analyse des correspondances* (T. II), Dunod, Paris.

Benzécri J.-P. (1981a): *Pratique de l'analyse des données*, T. 3, Linguistique & Lexicologie, Dunod, Paris.

Benzécri J.-P. (1991a): Typologies de textes grecs d'après les occurrences des formes des mots-outil, *Les Cahiers de l'Analyse des Données*, XVI, 1, 61-86.

Benzécri J.-P. (1991b): Typologies de textes latins d'après les occurrences des formes des mots-outil, *Les Cahiers de l'Analyse des Données*, XVI, 4, 439-465.

Benzécri J.-P. (1992): *Correspondence Analysis Handbook*, (Transl: T.K. Gopalan), Marcel Dekker, New York.

Benzécri J.-P., Benzécri F. (1992): Typologie de textes espagnols de la littérature du Siècle d'Or d'après les occurrences des formes des mots-outil, *Les Cahiers de l'Analyse des Données*, XVII, 4, Dunod, Paris, 425-464.

Benzécri J.-P. (1982): Construction d'une classification ascendante hiérarchique par la recherche en chaîne de voisins réciproques, *Cahiers d'Analyse des Données*, 7, 209-218.

Berelson B., Lazarsfeld P.F. (1948): *The analysis of communications content*, University of Chicago and Columbia University, Chicago and New York.

Bergounioux A., Launay M.-F., Mouriaux R., Sueur J-P., Tournier M. (1982): *La parole syndicale*, P.U.F., Paris.

Biber D. (1988): *Variation across speech and writing*, Cambridge University Press, Cambridge.

Biber D. (1990): Using register-diversified corpora for general language study, *Computational Linguistics*, 19, 2, 219-241.

Biber D. (1995): *Dimensions of register variation*. Cambridge University Press, Cambridge.

Blosseville M.J., Hébrail G., Monteil M.G., Pénot N. (1992): Automatic document classification : natural language processing, statistical analysis and expert system techniques used together, *Proceedings of the ACM-SIGIR*, Copenhagen, 51-58.

Bock H.H. (1994): Classification and clustering : Problems for the future. In : *New Approaches in Classification and Data Analysis*, Diday E. *et al.* (Eds), Springer Verlag, Berlin, 3-24.

Bolasco S. (1992): Sur différentes stratégie dans une analyse des formes textuelles : Une expérimentation à partir de données d'enquête, *Jornades Internacionals d'Analisi de Dades Textuals*, UPC, Barcelona, 69-88.

Bouchaffra D., Rouault J. (1992): Solving the morphological ambiguities using a nonstationnary hidden Markov model, *AAAI Fall Symposium on Probabilistic Approach to Natural Language processing*, Cambridge, Massachusetts.

Bouroche J.M., Curvalle B. (1974): La recherche documentaire par voisinage, *R.A.I.R.O.*, V-1, 65-96.

Bradburn N., Sudman S., and associates (1979): *Improving Interview Method and Questionnaire Design*, Jossey Bass, San Francisco.

Brainerd B. (1974): *Weighing Evidence in Language and Literature. A Statistical Approach*, University of Toronto Press, Toronto.

Brill E. (1992): A simple rule-based part-of-speech tagger, in: *Proceedings of the Third Conference on Applied Natural Language Processing*. A.N.L.P.'92. Trento, 152-155.

Brown P.F., Cocke J., Della Pietra S.A., Della pietra V.J., Jelinek F., Lafferty J.D., Mercer R.L., Roossin P.S. (1990): A statistical approach to machine translation, *Computational Linguistics*, 16, 2, 79-85.

Brunet E. (1981): *Le vocabulaire français de 1789 à nos jours, d'après les données du Trésor de la langue française*, Slatkine-Champion, Genève-Paris.

Burrows J.F. (1992): Computers and the study of literature, in: *Computers and Written Texts*, Butler C. S. (Ed), Blackwell, Oxford, 167-204.

Burrows J.F. (1992): Not unless you ask nicely: The interpretative nexus between analysis and information, *Literary and Linguistic Computing*, 7, 91-109.

Burt C. (1950): The factorial analysis of qualitative data, *British J. of Stat. Psychol.*, 3, 3, 166-185.

Burtschy B., Lebart L. (1991): Contiguity analysis and projection pursuit, in: *Applied Stochastic Models and Data Analysis*, Gutierrez R. and Valderrama M.J. (Eds), World Scientific, 117-128.

Carré R., Dégremont J.F., Gross M., Pierrel J.M., Sabah G. (1991): *Langage humain et machine*, Presses du CNRS, Paris.

Cazes P. (1981): Note sur les éléments supplémentaires en analyse des correspondances, *Les Cahiers de l'Analyse des Données*, 1, 9-23; 2, 133-154.

Celeux G., Hébrail G., Mkhadri A., Suchard M. (1991): Reduction of a large scale and ill-conditioned statistical problem on textual data, in: *Applied Stochastic Models and Data Analysis, Proceedings of the 5th Symposium in ASMDA*, Gutierrez R. and Valderrama M.J. Eds,World Scientific, 129-137.

Charniak E. (1993): *Statistical Language Learning*, The MIT Press, Cambridge.

Church K.W., Hanks P. (1990): Word association norms, mutual information and lexicography, *Computational Linguistics*, 16, 22-29.

Cibois P. (1992): Eclairer le vocabulaire des questions ouvertes par les questions fermées : le tableau lexical des questions, *Bull. de Method. Sociol.*, 26, 24-54.

Cohen M. (1950): Sur la statistique linguistique, *Conférence de l'institut de linguistique de l'université de Paris*, IX, Année 1949, Klincksieck, Paris.

Cooper W., Gey F.C., Dabney D.P. (1992): Probabilistic retrieval based on staged logistic regression, *Proceedings of the 15th Int. ACM-SIGIR Conf. on Res. and Dev. in Information Retrieval*, Belkin N and al., (Eds), ACM Press, New York, 198-209.

Cormack R.M. (1971): A review of classification. *J. of Royal Statist. Society*, Serie A, 134, 3, 321-367.

Cutting D.R., Karger D.R., Pedersen J.O., Tukey J.W. (1992): Scatter / Gather : A cluster based approach to browsing large document collections, *Proceedings of the 15th Int. ACM-SIGIR Conf. on Res. and Dev., in Information Retrieval*, Belkin N. *et al.*, (Eds), ACM Press, New York, 318-329.

Cutting D.R., Kupiec J., Pedersen J.O., Sibun P. (1992): A practical part-of-speech tagger, *Proceedings of COLING-92*, Nantes, France.

Deerwester S., Dumais S.T., Furnas G.W., Landauer T.K., Harshman R. (1990): Indexing by latent semantic analysis, *J. of the Amer. Soc. for Information Science*, 41 (6), 391-407.

Demonet M., Geffroy A., Gouaze J., Lafon P., Mouillaud M., Tournier M. (1975): *Des tracts en Mai 68. Mesures de vocabulaire et de contenu*, Armand Colin et Presses de la Fondation Nat. des Sc. Pol., Paris.

Diday E. (1971): La méthode des nuées dynamiques, *Revue Stat. Appl.*, 19, 2, 19-34.

Eckart C., Young G. (1936): The approximation of one matrix by another of lower rank, *Psychometrika*, l, 211-218.

Efron B. (1982): *The Jacknife, the Bootstrap and other Resampling Plans*, SIAM, Philadelphia.

Efron B., Thisted R. (1976): Estimating the number of unseen species: how many words did Shakespeare know ?, *Biometrika*, 63, 435-437.

Elia A. (1995): Per una disambiguazione semi-automatica di sintagmi composti: i dizionari electronici lessico-grammaticali, in: *Ricerca Qualitativa e Computer*, Cipriano R., Bolasco S. (Eds.), Franco Angeli, Milano.

Ellegard A. (1962): A statistical method for determining authorship: the Junius Letters, 1769-1772, *Gothenburg Studies in English*, 13, University of Gothenburg.

Escofier B. (1978): Analyse factorielle et distances répondant au principe d'équivalence distributionnelle, *Revue de Statist. Appl.*, 26, 4, 29-37.

Escoufier Y. (1985): L'Analyse des correspondances, ses propriétés, ses extensions. *Bull. of the Int. Stat. Inst.*, 4, 28-2.

Estoup J. B. (1916): *Gammes sténographiques*, 4th edition, Paris. (cf. aussi Mandelbrot, 1961).

Fagan J. (1987): Experiments in automatic phrase indexing for document retrieval; An examination of syntactic and non syntactic methods, *Proceedings of the 10th annual international ACM - SIGIR*, 91-101.

Fiala P., Habert B., Lafon P., Pineira C. (1987): Des mots aux syntagmes. Figements et variations dans la résolution générale du congrès de la CGT de 1978, *Mots*, 14, p 47-87.

Fisher R. A. (1936): The use of multiple measurements in taxonomic problems, *Annals of Eugenics*, 7, 179-188.

Forgy E.W. (1965): Cluster analysis of multivariate data : efficiency versus interpretability of classifications. *Biometric Society Meetings*, Riverside, California (Abstract in : *Biometrics* 21, 3, 768).

Fowler R.H., Fowler W.A.L., Wilson B.A. (1991): Integrating query, thesaurus, and documents through a common visual representation, *Proceedings of the 14th Int. ACM Conf. on Res. and Dev. in Information Retrieval*, Bookstein A. *et al.*, (Eds), ACM Press, New York, 142-151.

Friedman J.H. (1989): Regularized discriminant analysis, *Journal of the American Statistical Association*, 84, 165-175.

Froeschl K.A. (1992): Semantic metadata : query processing and data aggregation, in: *Computational Statistics*, Dodge Y., Whittaker J. (Eds), Physica Verlag, Heidelberg, p 357-362.

Fuchs W. (1952): On the mathematical analysis of style, *Biometrika*, 39, 122-129.

Fuhr N., Pfeifer U. (1991): Combining model-oriented and description-oriented approaches for probabilistics indexing, *Proceedings of the 14th Int. ACM Conf. on Res. and Dev. in Information Retrieval*, Bookstein A. *et al.*, (Eds), ACM Press, New York, 46-56.

Furnas G.W., Deerwester S., Dumais S.T., Landauer T.K., Harshman R.A., Streeter L.A., Lochbaum K.E. (1988): Information retrieval using a singular value decomposition model of latent semantic structure, *Proceedings of the 14th Int. ACM Conf. on Res. and Dev. in Information Retrieval*, 465-480.

Gabriel K.R. (1971): The biplot graphical displays of matrices with application to principal component analysis, *Biometrika*, 58, 453-467.

Gale W.A., Church K.W., Yarowski D.A. (1992): A method for disambiguating word senses in a large corpus, *Computers and the Humanities*, 26, 414-439.

Geisser S. (1975): The predictive sample reuse method with applications,*Journal of the American Statistical Association*, 70, 320-328.

Gifi A. (1990): *Non Linear Multivariate Analysis*, J. Wiley, Chichester.

Goldstein M., Dillon W.R. (1978): *Discrete Discriminant Analysis*, J. Wiley, Chichester.

Good I.J., Toulmin G.H. (1956): The number of new species and the increase in population coverage when a sample is increased, *Biometrika*, 43, 45-63.

Goodman L.A. (1986): Some useful extensions of the usual correspondence analysis approach and the usual log-linear approach in the analysis of contingency tables, *International Statist. Review*, 54, 243-270.

Gordon A.D. (1981): *Classification : Methods for the Exploratory Analysis of Multivariate Data.* Chapman and Hall, London.

Gordon A.D. (1987): A review of hierarchical classification, *J. of Royal Statist.Soc.*, A, 150, 2, 119-137.

Gower J.C. (1966): Some distance properties of latent and vector methods used in multivariate analysis, *Biometrika*, 53, 325-328.

Gower J.C., Ross G. (1969): Minimum spanning trees and single linkage cluster analysis. *Appl. Statistics*, 18, 54-64.

Gower J.C., Banfield C.F. (1975): Goodness-of-fit criteria in cluster analysis and their empirical distributions, in: *Proceeding of the 8th Intern. Biometric Conf.*, Corsten L.C.A., Postelnicu T., (Eds), 347-361.

Gower J.C., Hand D.J. (1996): *Biplots.* Chapman and Hall, London.

Greenacre M. (1993): *Correspondence Analysis in Practice*, Academic Press, London.

Greenacre M. (1984): *Theory and Applications of Correspondence Analysis*, Academic Press, London.

Greenacre M., Blasius J. (Eds) (1994): *Correspondence Analysis in the Social Sciences.* Academic Press, London.

Gross M., Perrin D. (1989): *Electronic Dictionaries and Automata in Computational Linguistics.* Lecture Notes in Computer Science, Springer Verlag, Berlin.

Guttman L. (1941): The quantification of a class of attributes: a theory and method of a scale construction, in: *The prediction of personal adjustment*, P. Horst, (Ed), SSCR New York, 251 -264.

Guttman L. (1946): An approach for quantifying paired comparison and rank order. *Ann. Math. Stat.* , 17, 144-163.

Guttman L. (1950): The principal component of scale analysis, in: S. A. Stouffer et al. (Eds) *Measurement and Prediction.* Princeton University Press, Princeton.

Habbema. D.F., Hermans J., van den Broek K. (1974): A stepwise discriminant analysis program using density estimation, in *COMPSTAT 1974*, Bruckman G. (Ed), Physica Verlag, Vienna.

Habert B., Salem A. (1995): L'utilisation de catégorisations multiples pour l'analyse quantitative de données textuelles, *Traitement Automatique des Langues*, 36, 1-2, 249-275.

Hand D.J. (1986): *Discrimination and Classification*, J. Wiley, New-York.

Hand D.J. (1992): Microdata, macrodata, and metadata, *Computational Statistics*, Dodge Y., Whittaker J. (Eds), Physica Verlag, Heidelberg, 325-340.

Hand D.J. (1996): Classification and computers: shifting the focus, in: *COMPSTAT 1996, Proceedings in Computational Statistics*, A. Prat (Ed) Physica Verlag, Heidelberg, 77-88.

Hardy A. (1994): An examination of procedures for determining the number of clusters in a data set, in: *New Approaches in Classification and Data Analysis*, Diday E. *et al.* (Eds), Springer Verlag, Berlin, 178-185.

Hartigan J.A. (1975) *Clustering Algorithms.* J. Wiley, New York.

Hayashi C. (1956): Theory and examples of quantification (II), *Proc. of the Institute of Stat. Math.* 4 (2) 19-30.

Hayashi C. (1987): Statistical Study of Japanese National Character, *J. Japan Statistical Soc.*, Special Issue, 71-95.

Hayashi C., Suzuki T., Sasaki M. (1992): *Data Analysis for Social Comparative research: International Perspective*, North-Holland, Amsterdam.

Hébrail G., Suchard M. (1990): Classifying documents: a discriminant analysis and an expert system work together, *COMPSTAT 90*, Momirovic K. and Midner (Eds), Physica Verlag, Heidelberg, 63-68.

Herdan G. (1964): *Quantitative Linguistics*, Butterworths, London.

Hirschfeld H.D. (1935): A Connection between correlation and contingency, *Proc. Camb. Phil. Soc.* 31, 520-524.

Holmes D.I. (1985): The analysis of literary style - A Review, *J. of Royal Statist. Soc.*, 148, 4, 328-341.

Holmes D.I. (1992): A Stylometric analysis of mormon scripture and related texts, *J. of Royal Statist.Soc.*, 155, 1, 91-120.

Holmes D.I. (1994): Authorship attribution, *Computers and the Humanities*, 28, 2, 87-106.

Holmes D.I, Forsyth R.S. (1995): The Federalists revisited: new directions in authorship attribution. *Literary and Linguistic Computing*, 10, 111-127.

Hotelling H. (1933): Analysis of a complex of statistical variables into principal components, *J. Educ. Psy.*, 24, 417-441; 498-520.

Jambu M. (1978): *Classification automatique pour l'analyse des données, Tome 1 : méthodes et algorithmes*, Dunod, Paris.

Juan J. (1982): Programme de classification hiérarchique par l'algorithme de la recherche en chaîne des voisins réciproques, *Les Cahiers de l'Analyse des Données*, VII, 2, 219-226.

Kaufman L., Rousseeuw P.J. (1990): *Finding Groups in Data*. J. Wiley, New York.

Labbé D. (1990): *Le vocabulaire de François Mitterand*, Presses de la Fond. Nat. des Sciences Politiques, Paris.

Lachenbruch P.A., Mickey M.R. (1968): Estimation of error rate in discriminant analysis, *Technometrics*, 10, 1-11.

Lafon P. (1980): Sur la variabilité de la fréquence des formes dans un corpus, *Mots* , 1, 127-165.

Lafon P. (1981): *Dépouillements et statistiques en lexicométrie*, Slatkine-Champion, 1984, Paris.

Lafon P., Salem A. (1983): L'inventaire des segments répétés d'un texte, *Mots 6*, 161-177.

Lance G.N., Williams W.T. (1967): A general theory of classification sorting strategies. *Computer J.*, 9, 373-380.

Lazarsfeld P.E. (1944): The controversy over detailed interviews - an offer for negotiation, *Public Opinion Quat.* 8, 38-60.

Lebart L. (1982a): Exploratory analysis of large sparse matrices, with application to textual data, *COMPSTAT*, Physica Verlag, Vienna, 67-76.

Lebart L. (1982b): L'Analyse statistique des réponses libres dans les enquêtes socio-économiques, *Consommation*, 1, Dunod, 39-62.

Lebart L. (1984): Correspondence analysis of graph structure, *Bulletin technique du CESIA*, 2, 5-19.

Lebart L. (1992): Discrimination through the regularized nearest cluster method, in: *Computational Statistics*, Y. Dodge, J. Whittaker, (Eds), Physica Verlag, Heidelberg, 103-118.

Lebart L. (1995): Assessing and comparing patterns in multivariate analysis, in *Data Science and application*, Hayashi *et al.* (Eds), Academic Press, Tokyo, Japan.

Lebart L., Mirkin B. (1993): Correspondence analysis and classification, in : *Multivariate Analysis: Future Directions 2*, C. M. Cuadras and C.M.Rao, (Eds), North Holland, Amsterdam, 341-357.

Lebart L., Morineau A., Warwick K. (1984): *Multivariate Descriptive Statistical Analysis*, J. Wiley, New York.

Lebart L., Salem A. (1994): *Statistique Textuelle*, Dunod, Paris.

Lebart L., Salem A., Berry E. (1991): Recent development in the statistical processing of textual data, *Applied Stoch. Model and Data Analysis*, 7, 47-62.

Lelu A. (1991): From data analysis to neural networks : new prospects for efficient browsing through databases, *Journal of Information Science*, vol. 17, 1-12.

Lewis D.D., Croft W.B. (1990): Term clustering of syntactic phrases, *Proceedings of the 13th Int. ACM Conf. on Res. and Dev. in Information Retrieval*, Vidick J.L., (Ed), A.C.M.Press, New York, 385-395.

Lewis D.D. (1992): An evaluation of phrasal and clustered representation on a text categorization task, in: *Proceedings of the 15th Int. ACM-SIGIR Conf. on Res. and Dev. in Information Retrieval*, Belkin N. *et al.*, (Eds), ACM Press, New York, 37-50.

MacQueen J.B. (1967): Some methods for classification and analysis of multivariate observations, in: *Proc. Symp. Math. Statist. and Probability (5th)*, Berkeley, 1, 281-297, Univ. of Calif. Press, Berkeley.

Mahalanobis P.C. (1936): On the generalized distance in statistics, *Proc. Nat. Inst. Sci. India*, 12, 49-55.

Mandelbrot B. (1961): On the theory of words frequency and on related Markovian models of discourses, Proc. of Symposia in Applied Math., XII, *Structure of Language and its Mathematical Aspects, Am. Math. Soc.*, Providence, R.I., 190-219.

Mandelbrot B. (1968): Les constantes chiffrées du discours, *Le Langage*, Encyclopédie de la Pléiade, XXV, Gallimard, Paris.

Margulis E.L. (1992): N-Poisson document modelling, *Proceedings of the 15th Int. ACM-SIGIR Conf. on Res. and Dev. in Information Retrieval*, Belkin N *et al.*, (Ed), ACM Press, New York, 177-189.

McKevitt P., Partridge D., Wilks Y. (1992): Approaches to natural language discourse processing, *Artificial Intelligence Review*, 6, 333-364.

McKinnon A., Webster R. (1971): A method of "author" identification, *The Comput. in Liter. and Linguist. Res.*, R.A. Wisbey, (Ed), Cambridge Univ. Press.

McLachlan G.J. (1992): *Discriminant Analysis and Statistical Pattern Recognition*, J. Wiley, New York.

McQuitty L.L. (1966): Single and multiple classification by reciprocal pairs and rank order type. *Educational Psychology Measurements*. 26, 253-265.

Mead R. (1988): *The Design of Experiments: Statistical Principles for Practical Applications*. Cambridge University Press, Cambridge.

Mosteller F. , Wallace D.L. (1964): *Inference and disputed Authorship : The Federalists*. Addison-Wesley, Reading, Mass.

Mosteller F., Wallace D.L. (1984): *Applied Bayesian and Classical Inference, the Case of the Federalist Papers*, Springer Verlag, New York.

Muller C. (1964): *Essai de statistique lexicale : L'illusion comique de P. Corneille*, Klincksieck, Paris.

Muller C. (1968): *Initiation à la Statistique Linguistique*, Larousse, Paris.

Muller C. (1977): *Principes et méthodes de statistique lexicale*, Hachette, Paris.

Nishisato S. (1980): *Analysis of Categorical Data. Dual Scaling and its Application*, Univ. of Toronto Press, Toronto.

O'Neill R., Wetherhill G.G. (1971): The present state of multiple comparison methods. *J. of Royal Statist. Soc.*, B, 33, 218-241.

Oostdijk N. (1991): *Corpus Linguistics and the Automatic Analysis of English*, Rodopi, Amsterdam.

Palermo D., Jenkins J. (1964): *Word Association Norm*, University of Minnesota Press, Minneapolis.

Partridge D. (1991): *A New Guide to Artificial Intelligence*, Ablex Publishing Corporation, Norwood, N.J.

Pearson K. (1901): On lines and planes of closest fit to systems of points in space, *Phil. Mag.* 2, 11, 559-572.

Pêcheux M. (1969): *Analyse automatique du discours*, Dunod, Paris.

Rao C.R. (1989): *Statistics and Truth*, International Cooperative Publishing House, Fairland, USA.

Reinert M. (1983): Une méthode de classification descendante hiérarchique: Application à l'analyse lexicale par contexte, *Les Cahiers de l'Analyse des Données*, 3, Dunod, 187-198.

Rugg D. (1941): Experiments in wording questions, *Public Opinion Quarterly*, 5, 91-92.

Salem A. (1984): La typologie des segments répétés dans un corpus, fondée sur l'analyse d'un tableau croisant mots et textes, *Les Cahiers de l'Analyse des Données*, 9, 4, 489-500.

Salem A. (1986): Segments répétés et analyse statistique des données textuelles, Etude quantitative à propos du père Duchesne de Hébert, *Histoire & Mesure, Vol. I-* 2, Presses du CNRS, Paris.

Salem A. (1987): *Pratique des segments répétés, Essai de satistique textuelle*, Klincksieck, Paris.

Salem A. (1993): *Méthodes de la statistique textuelle*, Thèse d'Etat, Université de la Sorbonne nouvelle (Paris 3).

Salton G. (1988): *Automatic Text Processing : the Transformation, Analysis and Retrieval of Information by Computer*, Addison-Wesley, New York.

Salton G., Mc Gill M.J. (1983): *Introduction to Modern Information Retrieval*, International Student Edition.

Sasaki M., Suzuki T. (1989): New directions in the study of general social attitudes : trends and cross-national perspectives, *Behaviormetrika*, 26, 9-30.

Saussure F. de (1915): *Cours de linguistique générale*, (reprinted:1986), Payot, Genève.

Schank, R.C., (Ed): *Conceptual Information Processing*, North-Holland, Amsterdam, 1975.

Schuman H. (1966): The random probe: a technique for evaluating the validity of closed questions, *Amer. Socio. Rev.*, 21, 218-222.

Schuman H., Presser F. (1981): *Question and Answers in Attitude Surveys*, Academic Press, New York.

Shannon C.E. (1948): A mathematical theory of communication, *B.S.T.J.*, 27.

Silberztein M. (1993): *Dictionnaires Electroniques et Analyse Automatique de Textes. Le Système INTEX*, Masson, Paris.

Simpson E.H. (1949): Measurement of diversity, *Nature*, 163, 688.

Smith M. W. A. (1983): Recent experience and new developments of methods for the determination of authorship, *Ass. for Lit. and Linguist. Comput. Bull.*, 11, 73-82.

Sneath P.H.A. (1957): The Application of computers to taxonomy. *J. General Microbiology*, 17, 201-226.

Sokal R.R., Sneath P.H.A. (1963): *Principles of Numerical Taxonomy*, Freeman and co., San-Francisco.

Somers H.H. (1966): Statistical methods in literary analysis, in: *The Computer and Literary Style*, J. Leed (Ed), Kent State University Press, Kent, Ohio.

Spevack M. (1968): *A Complete and Systematic Concordance of the Work of Shakespeare*, George Olms, Hildesheim.

Stone C.J. (1974): Cross-validatory choice and assessment of statistical predictions, *J. of Royal Statist. Soc., B 36, 111-147.*

Sudman S., Bradburn N. (1974): *Response Effects in Survey*, Aldine, Chicago.

Suzuki T. (1989): Cultural link analysis : its application to social attitudes - A study among five nations, *Bull. of the Int. Statist. Inst.*, 47th Session, 1, 363-379.

Tenenhaus M., Young F.W. (1985): An analysis and synthesis of multiple correspondence analysis, optimal scaling, dual scaling, homogeneity analysis and other methods for quantifying categorical multivariate data, *Psychometrika*, 50, 91-119.

Thisted R., Efron B. (1987): Did Shakespeare write a newly discovered poem?, *Biometrika*, 74, 445-455.

Thorndike R.L. (1953): Who belongs in the family, *Psychometrika*, 18, p 267-276.

Tournier M. (1980): D'ou viennent les fréquences de vocabulaire?, *Mots*, *1*, 189-212.

Tweedie F.J., Singh S., Holmes D.I. (1996): Neural network application in stylometry: The Federalist Papers, *Computers and the Humanities*, 30, 1, 1-10.

Van Rijckevorsel J. (1987): *The application of fuzzy coding and horseshoes in multiple correspondances analysis*, DSWO Press, Leiden.

Van Rijsbergen C.J. (1980): *Information Retrieval*, 2nd Ed., Butterworths, London.

Ward J.H. (1963): Hierarchical grouping to optimize an objective function, *Journal of the American Statistical Association*, 58, 236-244.

Weber R.P. (1985): *Basic Content Analysis*, Sage, Beverly Hills.

Weischedel R., Meteer M., Schwartz R., Ramshaw L., Palmucci J. (1993): Coping with ambiguity and unknown words through probabilistic models, *Computational Linguistics*, 19, 2, 361-382.

Wilks Y., Fass D., Guo C., McDonald J.E., Plate T., Slator B.M. (1991): Providing machine tractable dictionnary tools, *Machine translation*, 99-154.

Wilks Y., Guthrie L., Guthrie J., Cowie C. (1992): Combining weak methods in large-scale text processing, in: *Text based intelligent systems: Current Research and Practice in Information Extraction and Retrieval*, Jacobs P. (Ed), Lawrence Elbaum Associates, Hillsdale, (N.J.).

Wold S. (1976): Pattern recognition by means of disjoint principal component models, *Pattern Recognition*, 8, 127-139.

Wong M.A. (1982): A hybrid clustering method for identifying high density clusters, *Journal of theAmerican Statist. Assoc.*, 77, 841-847.

Wong M.A. (1985): A bootstrap testing procedure for investigating the number of subpopulations, *J. Statist. Comput. and Simul.*, 22, 99-112.

Yang Y. (1995): Noise reduction in a statistical approach to text categorization, in: *ACM / SIGIR'95*, Fox E.A., Ingwersen P., Fidel R., (Eds), Seattle, WA, USA, 256-263.

Yule G.U. (1944): *The Statistical Study of Literary Vocabulary*, Cambridge University Press, Reprinted in 1968 by Archon Books, Hamden, Connecticut.

Zipf G.K. (1935): *The Psychobiology of Language, an Introduction to Dynamic Philology*, Houghton-Mifflin, Boston.

Author Index

A

B

C

D

E

F

G

H

J

K

L

M

N-O

P

R

S

Subject Index

N-O

P

Q

R

S

Symbols

T	size of a corpus (number of occurrences)
V	vocabulary of a corpus (number of distinct words)
V_i	number of words whose frequency is i
F_{max}	frequency of the most frequent word in a corpus
k_{ij}	frequency of word i in part j of a corpus
$\mathbf{K}$	contingency table whose general term is k_{ij}
k	sum of the elements of the matrix $\mathbf{K}$ ($k = \sum_{i,j} k_{ij}$)
f_{ij}	relative frequency corresponding to k_{ij} ($f_{ij} = k_{ij} / k$)
ψ_i	coordinate of a row-point i on a principal axis
φ_j	coordinate of a column-point j on a principal axis
$\psi^+{}_i$	coordinate of a supplementary row i on a principal axis
$\varphi^+{}_j$	coordinate of a supplementary column j on a principal axis
λ_α	eigenvalue corresponding to axis α
τ_α	percentage of variance corresponding to axis α
R^n	n-dimensional space
+Exx	positive *characteristic element diagnostic* (ced) to which is associated a probability of the order $1/10^{xx}$
–Exx	negative *characteristic element diagnostic* (ced) to which is associated a probability of the order $1/10^{xx}$
↑Exx	positive characteristic *increment* to which is associated a probability of the order $1/10^{xx}$
↓Exx	negative characteristic *increment* to which is associated a probability of the order $1/10^{xx}$

Text, Speech and Language Technology

1. H. Bunt and M. Tomita (eds.): *Recent Advances in Parsing Technology.* 1996 ISBN 0-7923-4152-X
2. S. Young and G. Bloothooft (eds.): *Corpus-Based Methods in Language and Speech Processing.* 1997 ISBN 0-7923-4463-4
3. T. Dutoit: *An Introduction to Text-to-Speech Synthesis.* 1997 ISBN 0-7923-4498-7
4. L. Lebart, A. Salem and L. Berry: *Exploring Textual Data.* 1998 ISBN 0-7923-4840-0

KLUWER ACADEMIC PUBLISHERS – DORDRECHT / BOSTON / LONDON

GPSR Compliance

The European Union's (EU) General Product Safety Regulation (GPSR) is a set of rules that requires consumer products to be safe and our obligations to ensure this.

If you have any concerns about our products, you can contact us on ProductSafety@springernature.com

In case Publisher is established outside the EU, the EU authorized representative is:

Springer Nature Customer Service Center GmbH
Europaplatz 3
69115 Heidelberg, Germany

Batch number: 10584469

Printed by Printforce, the Netherlands